CRC Press
Taylor & Francis Group
6000 Broken Sound Parkway NW, Suite 300
Boca Raton, FL 33487-2742

International Standard Book Number-13: 978-1-138-70015-4 (Hardback)
International Standard Book Number-13: 978-1-315-20490-1 (ebook)

Library of Congress Cataloging-in-Publication Data

Names: Rantanen, Kalevi, author. | Conley, David W., author. | Domb, Ellen, author.
Title: Simplified TRIZ : new problem solving application for technical and business professionals /
Kalevi Rantanen, David W. Conley, Ellen R. Domb.
Description: 3rd edition. | Boca Raton : Taylor & Francis, CRC Press, 2018. | Includes bibliographical
references and index.
Identifiers: LCCN 2017014563| ISBN 9781138700154 (hardback : alk. paper) | ISBN 9781315204901 (eBook)
Subjects: LCSH: TRIZ theory. | Problem solving--Methodology. | Creative thinking. |Technological innovations.
Classification: LCC TA153 .R26 2018 | DDC 620.0076--dc23
LC record available at https://lccn.loc.gov/2017014563

Visit the Taylor & Francis Web site at
http://www.taylorandfrancis.com

and the CRC Press Web site at
http://www.crcpress.com

Contents

Preface to the Third Edition

People who need better tools for developing new systems, solving problems, and selecting solutions include:

- Those doing research and development in technical and nontechnical fields
- Technology managers and other managers
- Those who solve problems

The book presents a new model for problem solving. The model is based on the theory of problem solving called TRIZ. TRIZ is showing up more and more frequently today in success stories of innovative solutions to problems in technical and in nontechnical fields.

With these words, we began the first edition of *Simplified TRIZ* 15 years ago. Now they are more telling than ever.

There are five reasons to publish the third, completely revised and updated, edition of *Simplified TRIZ*.

First, copious experience has been collected thanks to the unprecedented growth of TRIZ all over the world.[1] We see the rise of TRIZ in China, India, and South Korea, and throughout Asia, while the progress is remaining steady in Europe and the Americas.

The third edition:

- Reflects a huge body of experience
- Contains substantial new material, including the following:
  - Three new chapters:
    - Important chapter on functional analysis
    - Chapter sharing two detailed case studies taking us from challenging problems to innovative solutions
    - Chapter sharing how to practice TRIZ tools "on the fly"
  - Multiple new case studies throughout
  - The addition of Lean to Chapter 16—Integrating TRIZ with Six Sigma and Other Quality Improvement Systems
  - More links between chapters, increasing the understanding of application
  - More application examples demonstrating application techniques of professionals
  - Examples of using the patterns of system evolution in real-world applications

In short, the third edition is a major movement forward in the world of teaching, utilizing, and experiencing TRIZ.

Second, TRIZ is being used more widely in general and much more in nontechnical fields. The subtitle has been changed: New Problem-Solving Applications for Technical and Business Professionals.

Third, TRIZ is increasingly being used in combination with other tools. At the same time, the need to find better tools for innovation has increased. We face difficult problems in our work. We need tools to help us focus on the correct issues and ignore false paths.

Fourth, there is more competition in the world economy. New tools of innovation are needed. There are more and more TRIZ users both in industry and academy.

Fifth, TRIZ-related research is increasing. There is more research inside the TRIZ community as well as in universities.

Many Fortune Global 500 companies such as Samsung, Intel, Ford, Procter & Gamble, and Mitsubishi have used TRIZ to develop better products more quickly. More and more small companies and individual inventors are using TRIZ. People in fields as diverse as marketing, education, and management are using TRIZ methods to solve their problems. People who thought that creativity was a trait that some have and some lack have learned that through the TRIZ process everyone can be creative.

Many users of well-known improvement methods such as Six Sigma, Quality Function Deployment, Taguchi methods, Design for Manufacturing and Assembly (DFMA), Lean, and others, have found that TRIZ is a valuable complement to the systems and methods they are using. TRIZ helps them use the other methods more effectively.

While big changes and continuous improvements are made, the strong and valuable traditions of TRIZ are maintained.

TRIZ comes from the Russian phrase *teorija rezhenija izobretatelskih zadach*. The English translation is the "theory of inventive problem solving." Why are people all over the world using a method with a Russian name?

TRIZ is an unusually global theory. It has a much wider multinational basis than most methodologies of management and creativity. The popularity of TRIZ in the United States, China, India, Japan, Germany, the United Kingdom, France, Korea, Israel, and other leading industrial countries is not a surprise, as the use of innovation as a business strategy has been strong for at least the past two decades. (See the works of Hamel[2] and Porter.[3,4])

TRIZ has its origins in the former Soviet Union, where it was founded by Genrich Saulovich Altshuller (1926–1998). He lived most of his lifetime in Baku, except for the years 1950–1954, when he was confined in prisons and camps, and spent his last years in Petrozavodsk, northern Russia. In research on patents that started at the end of World War II, Altshuller found that a variety of different engineering systems and technologies had common patterns of evolution. The inventor (or any problem solver) can learn these patterns of evolution, use them to develop new technology consciously and systematically, and avoid many fruitless trials and errors. His ideas about improving the way work was done did not please Stalin's administration, and he was arrested in 1950. In 1954, after Stalin's death, he was "rehabilitated" and returned to Baku. He continued to work on the new theory with a wide circle of colleagues. He conducted seminars and courses, mainly for engineers, and wrote books and articles. Over the next 30 years, the TRIZ methods developed.

Very little was known of this work outside the Soviet Union for more than 40 years. In the early 1990s, it became possible for Soviet experts to travel abroad and for some to emigrate. A TRIZ boom rose in the United States, Japan, and many other countries, in part because of the availability of software to support the use of TRIZ in English. Many new users, researchers, and service providers became involved. TRIZ became global. "Russian stuff" was mixed with customer-oriented approaches in the West. The result has been an extremely fruitful combination of Eastern and Western traditions.

TRIZ is the name of the theory, not a trademark. As TRIZ has proliferated, very different and sometimes contradictory things have been presented under the label "TRIZ." This book is different.

If we compare TRIZ with the automobile, this book is for drivers. You do not need to become an automotive engineer or mechanic to drive the car. You do not need to become an expert in the methodology to use TRIZ. If people tell you that TRIZ is complex, do not get worried and do not believe them. This kind of criticism probably reflects experience with the old TRIZ. At the birth of the automotive industry, drivers needed to be their own mechanics. At the birth of the computer era, only programmers could use computers. TRIZ has followed this pattern—the first TRIZ users were TRIZ experts. The situation is different today. Now everyone can learn TRIZ and use it effectively very quickly.

This book presents modern, international TRIZ. It is not a translation or review of older Russian books. The core concepts, which have been selected from the work of Altshuller and his colleagues, are as follows:

- Contradiction
- Resources
- Ideality
- Patterns of evolution
- Innovative principles

These concepts have passed the difficult tests of the market. The authors have tested them with their students and in their consulting work. They can be understood and used rapidly by beginners, and they are valuable to experts as well.

On the other hand, much material traditionally included in TRIZ texts has been left out. Long step-by-step guides or algorithms are avoided. Simple models have replaced some outdated and unnecessarily complex ones. Many good and interesting things traditionally included in TRIZ books have been left out because they are not useful to people who want to start using TRIZ quickly. If you need information not included in this book, you can find it easily by consulting the references provided in each chapter, especially from *The TRIZ Journal*,[5] which is a free online resource.

This book is a practical guide. It is a how-to book. It shows you and tells you how to solve problems creatively and—this may be even more important—shows you how to find problems and foresee the evolution of both the problem and the solution. The book contains many exercises, worksheets, and tables. You can download blank copies from http://www.innomationcorp.com/simplifiedTRIZ/.

We introduce case studies of different content and complexity. You can skip cases irrelevant to you.

At the same time, this is a strongly scientific work. The basic concepts and the models connecting them are emphasized. The word "scientific" means that the readers themselves should test and refine the generic model. We advise each reader to

- Not accept TRIZ only because it is in fashion or because well-known corporations support it
- Accept and embrace TRIZ because it works for you and for your problems

The results of behavioral sciences, especially findings of activity theory and cognitive psychology, have been used to develop this book. Its structure is designed to guide the reader through a successful learning and implementation process. Studying the book will take you through six stages:

1. Motivation (Why do I need new tools?)
2. Orientation (forming a general vision or mental model)
3. Internalization (enrichment of the mental model or getting new knowledge)
4. Application of the model to your own problems
5. Evaluation (testing and refining the model against your own experience)
6. Implementation (modifying the general process to work in your environment)

After studying this book, you will be eager to tell other people in your organization about TRIZ. To convince others, you should first convince yourself. For that, you need your own examples and cases. If you complete the exercises throughout the book, you will have a set of examples based on your own work. You can use each chapter and learn each tool separately, or you can work with the entire system. You will find that each tool is independently useful, but when used together, the system is even more helpful. Chapter 15 provides a road map for how to introduce TRIZ into your organization. Doing the exercises will help you help others to appreciate the need for TRIZ, which is the first step toward implementation.

The authors invite readers to send us their case studies, questions, and suggestions for use in future editions of *Simplified TRIZ*. We can be contacted at http://www.innomationcorp.com/simplifiedTRIZ/.

References

1. *Proceedings of the 12th MATRIZ. TRIZfest-2016 International Conference*. July 28–30, 2016. Beijing, People's Republic of China.
2. Hamel, G. 2000. *Leading the Revolution*. Boston, MA: Harvard Business School Press.
3. Porter, M. E. 1998. *On Competition*. Boston, MA: Harvard Business School Press.
4. Porter, M. E. 2001. Strategy and the Internet. *Harvard Business Review*. March 2001. https://hbr.org/2001/03/strategy-and-the-internet.
5. *The TRIZ Journal*. http://www.triz-journal.com.

Acknowledgments

Thanks are due to Pekka Koivukunnas from Metso Paper Corporation, who has offered many valuable user comments during the preparation of this book, as has Veli-Pekka Lifländer from Espoo-Vantaa Institute of Technology. The late Timo Saraneva, a friend and colleague, has helped to develop the appropriate model, and Ralph Czerepinski, Tom Kling, and Gregg Motter of the Dow Chemical Company contributed to our knowledge of how to teach TRIZ to people with a variety of different interests. Alan Conley, Cristen Conley, and Evan Davis, of Sunlight Homes, provided valuable input to much of the new third edition materials improving the accessibility of the knowledge to a wider audience. Finally, we acknowledge the contributions of Phil Samuel and Dan Laux and their colleagues at the Six Sigma Academy, who are pioneering the application of TRIZ in the Six Sigma process.

We would also like to thank our spouses, Galina Rantanen, Bill Domb, and Carole Conley, for their patience and support and many creative suggestions throughout the process of developing *Simplified TRIZ*.

Authors

Kalevi Rantanen is a Finnish TRIZ expert who successfully combines many different experiences and areas of knowledge in his work. In the 1970s, he worked with youth organizations, mainly on the problems of education and training. In the early 1980s, he studied in the former Soviet Union, earned his MS in mechanical engineering, and discovered for himself an unexpected, very exciting new world: TRIZ. He has worked in industry since 1985, and since 1991, he has been an independent entrepreneur. From 1991 to 2001, he has concentrated mainly on TRIZ training and, from 2002, on science and technology journalism.

David W. Conley is a TRIZ specialist who began his career as an Air Force Officer performing plasma physics and space nuclear propulsion research and served at Los Alamos and Brookhaven National Laboratories and on NASA's Nuclear Safety Review Panel. At Intel Corporation from 1995 until 2012, David held a variety of engineering and management roles and during his last 5 years with the company chaired Intel's worldwide innovation program, supporting the expansion and execution of TRIZ in the areas of product development, manufacturing process improvement, computing systems advancements, and business operations innovation. In 2013, he started Innomation Corporation and later joined the PQR Group as their Managing Partner. As a consultant, he has supported a wide base of organizations in contributing to quantum computing hardware development at Sandia National Laboratories, satellite superstructure and control systems at the Air Force Research Laboratory, crop breeding, pharmaceutical advancements, hospital operations, consumer electronics innovations, automotive design and polymer manufacturing, to name a few. He lives in New Mexico with his wife, Carole and sons, Dante, Roan, and Shane.

Ellen Domb is emeritus president of the PQR Group, a US consulting firm specializing in helping organizations maximize customer satisfaction, productivity, and profits through strategic management of quality and technology. Formerly a director of the Aerojet Electronic Systems Division with specific responsibility for total quality management implementation, she is a founding board member and judge for the California Council on Quality and Service. She is a charter member of the Quality Function Deployment Institute, cofounder of The TRIZ Institute, editor of *The TRIZ Journal*

(www.triz-journal.com) from 1996 to 2006, and chair of the first English language International TRIZ Symposium. Ellen is a popular speaker at TRIZ conferences worldwide, and has developed many techniques for training TRIZ trainers and incorporating TRIZ into quality improvement systems. Between trips she lives aboard a boat in Florida and scuba dives in the Bahamas with husband, Bill.

Chapter 1

Why Do People Seek New Ways to Solve Problems?

1.1 Introduction

In this book, we will study how to generate and select good solutions to problems using TRIZ, a new theory of problem solving. The term "TRIZ" (a Romanized acronym) comes from the Russian phrase *teorija rezhenija izobretatelskih zadach*, which means the "theory of inventive problem solving."

Why do we need a new theory? Without a theory, people generate ideas by guesswork and then select the ones they like or those they think other people will like. With TRIZ, you will be able to generate better ideas faster, and you will have a basis for selecting the best ideas, the ideas that will solve your problem effectively and form a basis for further improvements. In this chapter, we show that good ideas have frequently been rejected when first proposed. Much money, time, and human effort are lost when good ideas are rejected.

We show that people cannot select ideas properly and cannot produce better ideas effectively if they are unaware of the common features of good solutions: resolving contradictions, making use of idle resources, and increasing value. We show that common, traditional approaches to problem solving have often turned out to be dead ends. We need TRIZ, the theory based on the features of great inventions and the patterns of the evolution of systems, rather than approaches with no theory, based primarily on the emotions of the people involved.

1.2 Why Are Good Ideas Rejected?

People create new technologies and make creative use of existing technologies, generating many new ideas. How can we know which idea is good and which is not? History shows that companies and society have frequently rejected good ideas and invested money in ineffective ideas. Consider some examples:

- Alexander Fleming observed in 1928 that a mold culture produced something that was poisonous to many hazardous bacteria and not to humans. He named the new substance penicillin and published his results in 1929 in a well-known professional journal. In 1938, Ernst Chain read Fleming's article and became interested in penicillin. In 1939, he got a $5000 grant from the Rockefeller Foundation for the development of the new drug. It was the beginning of the penicillin industry. Why did scientists and investors ignore penicillin for 10 years? How was it possible that medicine and the drug business so long preferred older, often hopelessly ineffective, drugs and therapies to penicillin? Why did the world wait until the needs of World War II compelled it to seek seriously the new antibacterial drug?
- T-DRILL is a manufacturer of tube and pipe fabrication machines. The basic idea is simple: a collar is formed as part of the tube, from the material of the tube, replacing a conventional T-fitting. Only one joint is needed instead of three.[1] The benefits of T-joints without T-fittings in many applications are now indisputable. It took about 30 years to get this easy-to-understand idea accepted. Why?
- In 1948, Dick and Mac McDonald opened a fast-food restaurant where the customers themselves performed the function of waiters. In 1954, Ray Kroc looked at the McDonalds' stand. He saw that never had so many people been served so quickly. He understood the fast-food concept immediately. But why did it take 6 years for an entrepreneur to understand the concept? Why did the great majority of restaurateurs continue to offer poor service at higher prices?
- Molok is the trade name for a new bin for garbage and recyclables. The principle is simple. Molok is a vertical container, partly hidden underground—only 40% of the container is visible. The weight of the waste is used to compress the waste; you could say that gravity does the work. Because the new bin is partly underground, there is no odor. The container is lined with a big bag that can be removed and transported without complex specialized machinery. Why was this simple innovation not introduced until the 1990s? Why has it met considerable resistance?
- Flash smelting technology for copper, introduced by Outokumpu in the 1940s, is one of the most successful innovations in metallurgy in the twentieth century. Here, too, the general principle is simple. Sulfur contained in the ore enhances the fuel efficiency for smelting. The need for energy from outside is reduced drastically. Outokumpu's competitors also knew very well that if the ore contains sulfur, there is free energy available for smelting. Why did they ignore what they knew?

Every industry has examples like these. (Old: doctor or nurse does treatment. New: patient and family do treatment, professional advises. Old: teacher talks; student is quiet and listens. New: students learn by discussion with each other and teacher ....) What are some examples of good ideas from your industry, that were ignored when first introduced? Use the exercise Table 1.1 to collect examples of good solutions. Copies of workbooks and tables are available from www.innomationcorp.com/simplifiedTRIZ/.

> Why does it take years (or decades) for so many excellent solutions to be used, even though they are urgently needed and the technology is available?

This question has been asked many times in TRIZ classes, presentations, and discussions. The audiences usually offer some form of the following answers:

- The inventor is seldom a good salesperson.
- Lack of support from management.
- Poor presentation of the idea.
- Prejudices or the popular buzzwords "paradigm paralysis."
- NIH (not-invented-here) syndrome.

At first glance, the answers are self-evident. Closer examination reveals, however, that these answers do not help much. Imagine that all inventors become good salespeople, have the support of management, and have excellent presentation skills and materials. How can the inventor, product developer, or management know what idea is worth promoting? Companies have often used excellent sales skills to support outdated products. Richard Foster's book *Innovation*[2] gives many good historical examples. National Cash Register continued to advertise electromechanical cash registers in the 1970s when the development of electronics had already made them obsolete. The producers of cross-ply tires for cars were very customer oriented. This did not help when the superior steel-belted tire captured nearly the whole market in a short time. The list of examples can be easily continued: sailing ships versus steamships, vacuum tubes versus the transistor, pharmaceuticals versus homeopathics, conventional bike versus mountain bike, traditional passenger cars versus sport utility vehicles (SUVs), cabs and buses versus ride-sharing, universities versus at distance learning, suburbanization versus urbanization, fossil fuels versus alternative energies, and so on. The crucial point here is that content matters. It is trivial to say that inventors should be able to get their new ideas accepted. How can they know what idea is really new and better than the old technology?

What about prejudices? Would the result be better if experts and managers were less prejudiced and more creative? It is true that the inventor should be open to new ideas and criticism. Every idea needs a champion who can fight stubbornly against resistance and indifference. How can inventors know when to accept feedback or criticism and when to reject it? Again, the content of the idea is important. One must select a good solution from many ideas, some good and some

Table 1.1   Examples of Good Solutions

Year Created	Year Implemented	Idea
Blank templates can be downloaded from www.innomationcorp.com/simplifiedTRIZ/		

bad. How can we select the best solution? Is it best for our customer? Best for our business? Is it the most interesting technology?

We propose a simple reason for the rejection of good ideas: people reject good solutions and invest in bad ones because they do not know the difference between them.

Looking at cases of lost opportunities and great losses for business and society, people can take one of two positions:

1. With hindsight, it is easy to see that often very good ideas are rejected and resources lost to bad ones. Obviously, it is not possible to know whether the idea is good or bad when it is first proposed.
2. Because the same patterns are repeated, we can learn from the past. The patterns of evolution can be discovered and used to get better solutions today.

There is growing evidence supporting position 2, using TRIZ to provide the general theory of the evolution of good ideas. See, for example, the statement from a participant of a TRIZ class conducted by one of the authors (names are removed, the rest is cited verbatim):

> "A sad but true testimony to the power of TRIZ. In one of our TRIZ sessions that you conducted at XXX, we identified the use of water, transformed into steam, as a method for foaming an adhesive. This idea, though considered valid, was never acted on. A patent was recently issued to one of our competitors for a process of foaming an adhesive with water vapor."

1.3 Common Features of Good Solutions

Good solutions have several common features. The good idea does the following:

- Resolves contradictions
- Increases the "ideality" of the system
- Uses idle, easily available resources

In addition to their everyday meanings, these words have specific technical meanings in TRIZ. By working with the TRIZ concepts, you will learn to apply them to your problems, to develop good solutions, and to select the best solutions from all that are proposed.

1.3.1 Three Basic Concepts for Reaching the Best Solution

1. A good solution resolves the contradiction that is the cause of the problem. There are two kinds of contradictions:
   a. "Tradeoff contradiction" means that if something good happens, something bad happens, too.
   b. "Inherent contradiction" means that I need one thing that possesses two opposite properties.
2. The "ideality" of a system is the measure of how close it is to the perfect system. The perfect system (called the "Ideal Final Result" in TRIZ) has all the benefits the customer wants, at

no cost, with no harmful effects. So, a system increases in ideality when it gives you more of what you want or less of what you do not want, does it at a lower cost, and does so usually with less complexity.

3. Unseen, idle resources of the system are used to reach these seemingly incompatible goals. These resources include energy, materials, objects, information, or things that can be made easily from the resources that are in the system or nearby.

All five examples of resistance to new technology illustrate clearly these concepts of overcoming contradictions, increasing ideality, and using resources:

1. Penicillin resolved a typical contradiction of drugs: substances that can kill microbes destroy healthy tissues, too. Mold, present everywhere, was used for resolving contradiction. Increasing ideality: many important diseases, earlier considered totally hopeless, were easily cured by penicillin.
2. Collar made as part of the tube: there is no separate T-fitting. The tube is used as a resource. Ideality increases: one joint is less complicated, requires less material, and uses less labor than three.
3. Fast-food restaurant: there are no waiters, but, at the same time, all customers have their own waiter, that is, they serve themselves. Resource: a customer. Increasing ideality: better and quicker service.
4. A bin for garbage should be big and at the same time small. A partly underground container is big (in available volume) and small (the part you see). The space under the bin is an easily available resource.
5. Flash smelting resolves a contradiction: much energy is needed and energy should not be used at all. Sulfur in the ore is an easily available energy resource.

These three concepts are repeated. Contradictions are solved. Idle resources are used. Solving contradictions by using resources makes the system more ideal. We can describe the movement from the problem to the solution by a simple diagram (Figure 1.1).

We see that creative activity in research and development (R&D), product development, manufacturing, marketing, management, and other areas needs reorganization. A simple scheme (Figure 1.2) shows the desired changes.

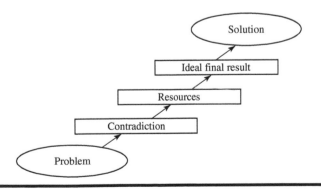

Figure 1.1 Features of good solutions. Contradictions are solved. Idle resources are used. Solving contradictions by using resources makes the system more ideal.

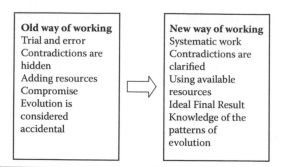

Figure 1.2   The reorganization of creative activity. The transition from the old way of working to the new.

1.4 A New Approach to Problem Solving Is Needed

There have been, of course, many attempts to make creative work more effective and to replace the trial-and-error method. These approaches have different names. However, they can be easily divided into two groups:

1. The first group can be called the "rationalized" or "hard" model. McGregor's Theory X describes this model well.[3] R&D centers are established. The work is controlled by budgets and time limits. The silent assumption is that people need to be controlled and directed rather tightly. This kind of management helps to get minor improvements, but seldom gives great, qualitatively new ideas.
2. Many attempts have been undertaken to overcome the weaknesses of the rationalized model. Many creative techniques have been offered. There are few substantive differences between them. Together they can be called a "humanized" or "soft" model, which fits McGregor's Theory Y. In this model, people naturally have imagination and creativity. External control is not the only means for getting good results. So criticism and control are minimized or prohibited. Fantasy, feeling, play, intuition, and pleasure are encouraged.

The humanized model is often attractive in the beginning. Many ideas are generated. Soon, however, most proposals turn out to be the repetition of old inventions. Sometimes, really good ideas are developed, but they are not recognized due to the lack of evaluation criteria (see examples in Section 1.2).

Christopher Freeman[4] has characterized demand-pull theory and science-push theory as two poles in the debate on the determinants of innovation. Rationalized activity and humanized activity are considered two opposite means to improve the traditional craft activity, as described by Engeström.[5] Theory X and demand-pull theory can be loosely related to the rationalized model, Theory Y and science-push theory to the humanized model.

Disappointment in the soft approach causes organizations to return to the hard model. Then, after some time, traditional management is criticized for the lack of creativity and free idea generation comes into fashion again. And so on. Inventive methodologies in industry seem to oscillate perpetually between hard and soft models. Both ways are blind alleys. Both are unsatisfactory.

There are many good strong tools for the development of systems, but they all seem to be missing specific techniques for problem solving. Such methods include the following:

- Quality Function Deployment (QFD): Identifies the voice of the customer and helps the organization understand where creative ideas are needed. However, it has no tools to create new concepts to meet the customers' often-contradictory requirements.
- Theory of Constraints (TOC): Helps to define conflicts and to identify where a conflict resolution is required, but it does not have tools and techniques for generating the ideas that will resolve the conflict.
- Six Sigma: In 10 years, it went from "just another quality system" to a corporate management system that gets Wall Street's attention for its ability to mobilize organizations. Six Sigma focuses on the reduction of variation and integrates many methods of problem identification and analysis, but it took until 2001 for some organizations that teach Six Sigma to begin incorporating TRIZ to get good solutions to the problems that were identified.
- Lean: Identifies seven types of waste but simply tells the user to "get rid of it" once it has been identified. The elimination of waste often results in contradictory requirements within the system being analyzed. Some organizations have started using TRIZ with Lean (like using TRIZ with Six Sigma) to help develop solutions around the waste that has been identified.

Chapters 15 and 16 of this book will deal with incorporating TRIZ into these methods so that organizations can combine the power of TRIZ for solving problems with the power of any of the methods for finding problems.

A new approach, neither hard nor soft, but incorporating the benefits of both views, has become necessary. It is not a mechanical sum of traditional ways to think. It is the TRIZ system of understanding the problem, modeling the contradictions, removing them by using resources, and improving the ideality of the system, not relying on intuition. It relies on knowledge of the system being improved and on knowledge of the systematic method for improvement.

TRIZ is based on more than 70 years of research, but it is somewhat new to most of the industrial world. Increasing consciousness of the weakness of traditional approaches has increased interest in TRIZ. TRIZ does *not* ask, "What is the difference between creative and uncreative people or organizations?" TRIZ asks, "What is the difference between good and bad *ideas,* solutions, and products?" TRIZ seeks the sources of creativity in the objects or systems to be improved in the outer world, not in the psychology of the people or the organization doing the work.

A simple comparison illustrates the approaches. Runners can increase their speed using physical and mental exercises. A coach can manage runners using the hard or soft way: compel them to do structured exercises or give them freedom to run however they want. Both methods have been used to increase speed and certain methods work better with certain runners. The speed can be increased, but not very much and not very quickly. A different way is to say that the goal is to go fast and to provide the runners with vehicles: bicycles, cars, planes, and boats. Now the main point is not differences between people, but differences between tools and everyone can go fast. TRIZ offers vehicles for moving to better ideas, solutions, and innovations. Knowledge of the features of good solutions is the vehicle that can be used to generate better solutions. The point is not to learn more about the psychology of people to increase creativity, but, using TRIZ, to develop creative ideas, no matter what kind of intuitive skills each has.

All four approaches we have considered are presented in a simple schematic drawing (Figure 1.3). These models describe a generic framework. If you model your own experience, it will give this framework more meaning for you. The exercises in Table 1.2 will help you do this. Start by filling in a description of the methods that your organization has used for problem solving and for stimulating innovation. If you have gone back and forth from one method to another, draw arrows on the table to show the path. If you have tried TRIZ or a method related to TRIZ,

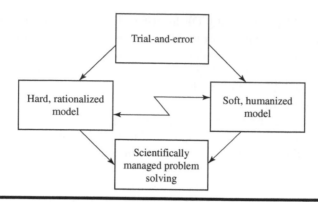

Figure 1.3   Models of creative work. Trial-and-error method is the oldest way of seeking ideas. Hard and soft models are two popular attempts to make work more effective. Scientifically managed problem solving combines the strengths of both approaches with the theory of TRIZ.

Table 1.2   Methods Used by Your Organization for Problem Solving and Stimulating Innovation

Trial-and-Error Methods	
Hard, Rationalized Model	*Soft, Humanized Model*
Scientifically Managed Problem Solving	

list it under Scientifically Managed Problem Solving and use arrows to show the path from other methods to TRIZ.

1.5 Summary

To generate and select ideas for good solutions to problems, one should know the difference between strong and weak ideas. To promote ideas, you should know which solutions are worthy of being promoted.

Traditional methods of problem solving do not have criteria for selecting good ideas.

A strong solution resolves a contradiction, makes use of idle resources, and increases the ideality of the system. TRIZ is the theory that provides the basis for this model of successful problem solving.

The next chapter gives an overview of the new model for problem solving.

References

1. T-DRILL. 2017. Tee forming, brazed joint, T-DRLL collaring method. http://www.t-drill.com/technologies/tee-forming/.
2. Foster, R. N. 1986. Innovation: The Attacker's Advantage. New York: Summit Books.
3. McGregor, D. 1960. *The Human Side of Enterprise*. New York: McGraw-Hill.
4. Freeman, C. 1979. The determinants of innovation. *Futures*, 3, 206–215.
5. Engeström, Y. 1987. *Learning by Expanding*. Helsinki: Orienta-Konsultit, 284.

Chapter 2

Constructing the New Model for Problem Solving: Moving from the Problem to the Ideal Final Result

2.1 Introduction

In the first chapter, we showed that a new approach to problem solving is needed and briefly outlined the basic features of TRIZ. In this chapter, we construct a model for problem solving. The model is like a general map that shows, by words and pictures, how to use the most important TRIZ features in problem solving. Details will be studied in later chapters. This short chapter is very important. The model presented will guide you through the details and help keep you on track as you study and use TRIZ.

We will construct the model in six steps. First, we describe the concept of contradiction. Second, mapping of resources is added to the model. Third, the concept of the Ideal Final Result is formulated. These steps form the inner shell of the model. The fourth, fifth, and sixth parts of the model are functional analysis, the patterns of evolution, and innovative principles.

This model for problem solving is based on the theory of TRIZ, on customer feedback from people who have used it, and on the knowledge of the styles of human thinking and problem-solving activity.

2.2 Contradiction

2.2.1 Difficult Problems Contain Contradictions

One of the early insights of the TRIZ researchers was that solving a problem meant removing a contradiction. If we compare the TRIZ problem-solving methodology to a tree, the concept of contradiction can be compared to the seed, from which we can grow the whole tree. If we would like to express the idea of TRIZ by a single word, that word would be "contradiction."

A contradiction is a conflict in the system. A system consists of at least two components: tool (*T*) and object (*O*). The ax, for example, is a tool that splits the object, a chunk of wood. Splitting power is a good feature that is connected with less desirable properties, such as the clumsiness of the tool. If you make the blade heavier, the ax can strike a more effective blow, but it becomes more awkward to handle.

We meet contradictions everywhere. For example, a company wants to improve business by improving customer service and decides to get better service by increasing staff training. Training is the *T* that is used to improve a certain *O*: employees (specifically their professional quality). If employees get extensive and thorough training, service surely can be improved, but the time loss might be intolerable. We get a contradiction: the better the service, the more training time is needed.

Think of a seesaw (Figure 2.1). When one end of the plank goes up, the other goes down. You cannot get both ends to go up at the same time.

In this case, the connection of features (up-down) is not a problem—having one end go up when the other goes down is a natural property of a seesaw. If you do not want the experience of going up and down, you will need to choose a different toy or make some major changes to the seesaw system.

The first three concepts can be illustrated by a simple figure (Figure 2.2) containing a tool, an object, and a contradiction.

We will use a diagram like this for each problem. A flash-like arrow between a tool and an object indicates a contradiction. Many visual, mathematical, and physical models are available for the design, use, and maintenance of systems. This simple model is easy to use in deciding whether you need TRIZ to solve the problem. Drawing the diagram makes you think about the problem and decide whether you have a contradiction. You can solve many kinds of problems using TRIZ, but the theory is most powerful and gives the most value added when used to solve nonroutine (i.e., the "inventive" in the "theory of inventive problem solving") problems containing contradictions.

Figure 2.1   The seesaw analogy of the problem situation.

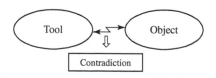

Figure 2.2   Contradictions between tools and objects are the moving force of evolution.

2.3 Resources

Sometimes, the clear formulation of the contradiction suggests a possible answer to the problem. Usually, however, additional information is needed. Resource analysis helps you find ways to resolve the contradiction.

Resources are things, information, energy, or properties of the materials that are already in, or near, the environment of the problem. If they can be used directly, or modified to make them useful, the problem will appear to have solved itself. Think of the resources as the reserves—they are invisible at first, because we are accustomed to not seeing them when we look at the problem situation, but we can mobilize these reserves to solve the problem. That is why we add the block "Resources" to our model (see Figure 2.3).

The system should be changed so that the needed improvement seems to appear from nowhere. For example, in Chapter 1 we described a garbage bin that was partly hidden underground. The needed change was achieved using the space beneath the bin.

In the ax example mentioned earlier, we need to change the system so that splitting power is improved, but the ax does not get more difficult to use. The resources in the ax and chunk of wood system are the blade, its edge, form, material, and other properties; the handle and its properties; the chunk of wood and its properties; the surrounding air; and so forth. Depending on how the system is defined, we can also include the person whose arm provides the energy and whose hands provide the transmission of energy to the ax.

In the training example, the curriculum, its structure and properties, the skills of teachers, textbooks, and the motivation and existing knowledge and skills of the students are all resources, as well as the culture of the company, the employee's work schedule and structure, and the physical environment of the training.

2.4 The Ideal Final Result

Using resources, one can remove the contradiction and get the Ideal Final Result. This is the third concept that will be added to the model (Figure 2.4).

Imagine that both ends of the plank forming the seesaw go up by virtue of resources. If the child's resources included some rope, a tree, and a strong parent, both ends of the plank could go up at the same time (Figure 2.5).

The blow struck by the ax should be made stronger, but, at the same time, the tool should remain easy to use. To split the chunk of wood, the ax should be heavy, but, at the same time, for ease of handling, the ax should be light.

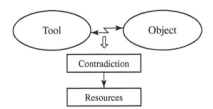

Figure 2.3 Resources are information, energy, properties, and such, available for solving contradictions. They are often invisible at first, because we are accustomed to not seeing them when we look at the problem situation.

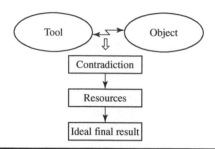

Figure 2.4   The Ideal Final Result is the solution that resolves the contradiction without compromise. Resources are used to go from the contradiction to as perfect a solution as possible.

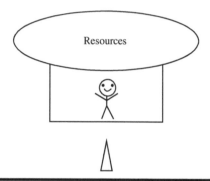

Figure 2.5   The seesaw analogy of the Ideal Final Result. Both ends of the plank go up.

The Ideal Final Result can now be described as follows: something changes the ax in some way so that it is both heavy and light, for the double purpose of making the blow stronger without decreasing the ease of use. The ax marketed by Fiskars solves this problem by using a hollow handle. Air is used as the resource that solves the problem. The hollow handle gives an unexpected new quality. The center of gravity moves nearer to the blade. The blow is more powerful, which could be thought of as heavy, although the tool is lighter.

In the ideal training, results are improved, but the length of time is not increased. Changing the training so that part of it is done on the job, not in the classroom, frequently results in better learning as well as shorter classes, because the students can see the results immediately and are in a better position to apply them. This improves both motivation and feedback.

"Ideality" is the measure of how close the system is to the Ideal Final Result. If the useful feature improves, the ideality improves. If the less than desirable feature decreases, the ideality also improves. To solve the problem, look for resources already in the system that can help the useful feature get better or the less than desirable feature be reduced—or vanish entirely.

2.5 Functional Analysis

Functional analysis is a method used to understand the relationships between the system's physical objects (components) by way of their functions. In the ax example, the arm *moves* the hand; the hand *holds* and *moves* the ax handle; the ax handle *holds* and *moves* the ax blade; and the ax blade *cuts* the wood. Functional analysis helps us to understand where there are unwanted (harmful), and inappropriate levels (insufficient and excessive) of, functions. Understanding where our system has unwanted and inappropriate levels of functions leads directly to understanding where contradictions exist within the system. Further, functional analysis supports the comprehension and inventorying of resources in and around the system's environment. More detail on functional analysis is provided in Chapter 7—Understanding How Systems Work: Utilizing Functional Analysis to Expand Knowledge About Your Problem.

2.6 Patterns of Evolution

Formulate the contradiction, map the resources, and define the Ideal Final Result. Is this enough? Sometimes yes. Often, however, something more is needed to move from the Ideal Final Result to the technical solution of the problem. We need methods to follow to resolve contradictions, to use resources, and to make the system more ideal. One method is the use of the patterns of evolution of systems. Features are, as in everyday language, any properties of a system: size, weight, speed, flexibility, color, and so on. Patterns of evolution are important regularities in system development, for example, the transition from the macro- to the microlevel or the division of the system into smaller parts. Patterns are actually laws, but they are soft, not rigorous mathematical formulas as in physics. That is why we usually refer to them as patterns.

Studies of the history of innovation have shown that many improvements follow similar patterns. We just named the transition to the microlevel or division of the system into multiple parts. Examples are all around us. In your computer and printer, electrons and other microscopic parts have replaced the cumbersome parts of old typewriters and calculators. In the kitchen, you have a microwave oven. The food is heated by electromagnetic vibration of its water molecules, not by radiant heating from a heavy metal plate or flame. You may work in an organization that has divided itself into many relatively independent teams. If you like outdoor activities, you can wear clothes made from microfibers. In your household work, you may use microfiber cloths for cleaning. The bed in your bedroom is also making the transition to microlevel. Water beds, air beds, or mattresses composed of small cells of some kind in place of steel innersprings are superior to traditional beds. Everything seems to get divided into smaller parts.

Different parts of a system may change at different rates and may follow different patterns. The evolution is uneven. We spoke of the transition to microlevel. There is the transition to macrolevel (also called the "supersystem"), too. The parts of the system become more interactive with each other. The system is expanded and convoluted. It is improved by adding more and more features, then combining all the features into a new, simpler system that has all the benefits without all the complexity.

We can use these patterns to find hints about how any situation could be improved and to obtain suggestions about how the system could be changed to become more ideal. Chapter 10 is a detailed presentation of how to use the patterns of evolution.

In the ax example, one can think of the transition to the microlevel. The ax can be segmented or divided into parts. If you continue segmenting it into smaller and smaller parts, eventually you get particles, then molecules. An ax made of molecules? A gaseous ax? Is this crazy? The hollow ax, containing air in the handle, is partially gaseous. Or, a stream of particles, as with a sandblaster, can be a very effective cutting tool, where each particle could be considered a micro-ax.

Some important patterns of evolution can be applied to business systems. Training is a system that evolves unevenly. A training program can be segmented (transition to microlevel) and integrated to larger systems (transition to macrolevel). Programs tend to expand and then inevitably be

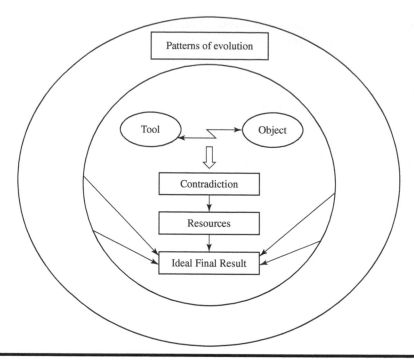

Figure 2.6   The patterns of evolution have multiple uses: they support the tools for problem solving, help to get solutions directly, and can be used for the prediction of the future features of technology.

compressed so that efficiency is maintained or increased. The simplified model of TRIZ presented here is an example of trimming a complex subject to make it easy for people to get started using it.

We add the patterns of evolution to the pictorial model. They can often be used directly to develop good solutions to problems, as well as to predict the future evolution of the system. Figure 2.6 suggests the use of multiple patterns of evolution. To keep the model simple, arrows are drawn only from the patterns of evolution to the Ideal Final Result. This connection is most important for problem solving because the more ideal system is what we need.

2.7 Innovative Principles across Industries

We now have five important concepts in the model: contradiction, resources, the Ideal Final Result, functional analysis, and the patterns of evolution. However, these are not always enough. The evolution pattern of the system may suggest a rather vague idea of solution, and we may need much more specific help. Innovative principles are tools that help us to understand what a pattern might mean and how a pattern might be applied for any particular problem. There are 40 Principles, which will be studied in detail in Chapter 11. A few examples show how they are used.

For the example of the improvement of the ax, it might not be obvious at first that the replacement of a solid body with a hollow one follows the pattern of transition to a microlevel. Principle 29 gives a more concrete hint: "Pneumatics and hydraulics. Use gas and liquid as parts of an object ..." Now, one needs to remember only that air is also gas to see that a part of the ax can be made from air. The idea of the hollow handle is generated nearly automatically.

The 40 Principles are based on the same study of patents and technology that developed the patterns of evolution. In the larger sense, this is a study of how people solve problems, not just a study of technology. That is why the same principles can frequently be applied to problems in management, marketing, training, and other fields. For example, Principle 18—Mechanical Vibration suggests using an object's resonant frequency. It may mean resonance or synchronization in mechanical, electromagnetic, or acoustical systems, but in training, it can also be interpreted as improving the coordination of the curriculum, textbooks, and teaching with the learning style of the students and the culture of the company.

Let us add principles to our model (see Figure 2.7). The arrow from the principles to the Ideal Final Result shows that these are shortcuts that sometimes allow us to bypass the analysis of contradictions and resources. However, using the whole model is more effective than using the parts separately. Innovative principles are studied in detail in Chapter 11.

2.8 Other Concepts and Tools

Figure 2.7 is the model that will be used throughout this book. It is easy to remember and easy to use. It gives you the power of TRIZ very quickly. Traditional TRIZ has many more tools that you may want to explore after mastering the tools and concepts presented in this book. Figure 2.8 shows an enhanced model that you can use if you study any of the other TRIZ tools later.

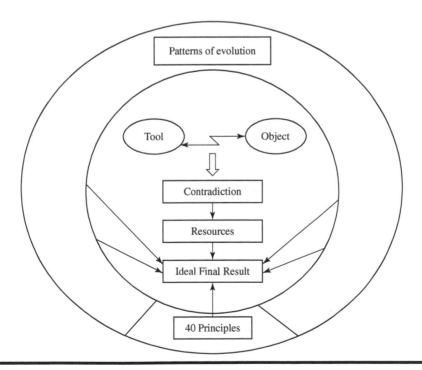

Figure 2.7 Forty innovative principles give cues for finding ideas. They can be used both as independent tools and to support other methods.

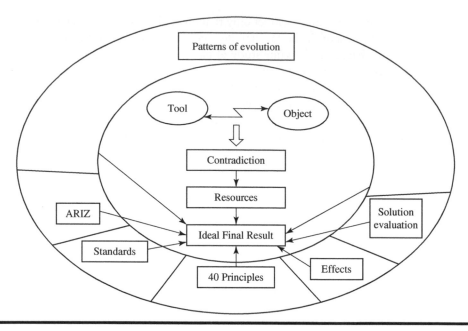

Figure 2.8   Many other tools can be added to the model.

The model also shows four other tools: ARIZ, standards, effects, and solution evaluation. ARIZ (algorithm for inventive problem solving) is a long step-by-step guide for the analysis and resolution of contradictions. Standard solutions list the ways of transforming the system, based on Altshuller's 1985 publication of a list called the "76 Standard Solutions." (One of them has 10 parts.) Effects is a technical database of physical, chemical, mechanical, biological, geometrical, and other technical phenomena that can be used for inventive problem solving. Various forms of the database appear in textbooks, software, and online resources. Standard solutions, principles, and effects are lists of recommendations and examples that can be easily captured within computer software. There are many TRIZ software offerings on the market. However, it is important to keep in mind that just as a calculator will not teach you how to do mathematics neither will TRIZ software teach you how to perform innovation analyses. There are many solution evaluation techniques that can be used to rate and order your solutions. As an example, solution concepts can be graded against quality (how well does the solution concept solve the problem without creating conflicts?), time (how long will it take to get the solution in place?), or investment (how much will it cost to implement the solution, including testing and training the people who will use it?). Alternatively, the solution can be evaluated in terms of TRIZ—the closer the solution is to the Ideal Final Result the better. Your business situation will determine which criteria should be used for the evaluation. Many other tools and techniques are available from other books and from online resources.

2.9 Why Introduce This Model?

First, the model presented in this book is based on modern TRIZ, on recent achievements of innovative design in industry, and on the latest research. The research on the evolution of technical, and other, systems guided the creation of the model and the organization of tools in the model.

At the same time, we have carefully kept intact some important old tools that have proven to be fruitful, handy, and robust during many years of application. The most famous of these classical tools is the list of 40 Principles presented in Chapter 11.

Second, the general concepts described by the model are based on feedback from users. Users we have met in our training classes prefer general concepts to long procedures. They have told us that they prefer to study concepts first, then study short step-by-step guides and checklists, and then apply the concepts to their own problems. They like the contradiction concept and the ideality concept as new lenses for seeing reality. They usually dislike long instructions—it seems that an instruction longer than one page will never be used.

Third, we use the results from behavioral sciences gained from the study of human activity. Activity theory and cognitive psychology have shown that individuals and teams need general organizing models to solve problems effectively. Work researchers use a model of human activity containing subject, tools, object (there is considerable overlap with the definition of a system in TRIZ), and also community, rules, and division of labor, see Engeström.[1] Peter Senge and his team speak of "mental models" and "shared visions" in the organization.

All kinds of models have become fashionable in recent years. Many different words are used: model, internal model, mental model, paradigm, vision, schema, and others. However, it is not enough to say that internal models are necessary. The model should adequately describe the essential features of the object. In his article on the history of the transistor, Shockley describes the foyer of the main entrance to Bell Laboratories, where the following statement credited to Alexander Graham Bell is posted: "Leave the beaten track occasionally and dive into the woods. You will be certain to find something that you have never seen before."[2]

The statement encourages traditional problem solving through the trial-and-error method. Though, through the usage of TRIZ, the trials result in much less error, as TRIZ is based on the successful innovation of those before you. The mental model of TRIZ can be compared to a map and a compass. One leaves the beaten track, but not with empty hands.

The model is not an arbitrary construction, but the reflection of the system that is the object of creative activity. The model as presented here will help beginners get a fast start in obtaining useful and creative results. As you learn more and more about TRIZ, you will modify the model and, as TRIZ research continues, more methods and tools will become available for inclusion in the model. The user is encouraged to test and improve the model continuously.

2.10 Summary

The problem-solving model uses six concepts: contradiction, resources, the Ideal Final Result, functional analysis, the patterns of evolution, and innovative principles.

1. Contradiction: Solving a problem means removing a contradiction. Contradictions are considered in detail in Chapters 3 and 4.
2. Resources: Resources are available, but idle and often invisible substances, energy, properties, and other things in or near the system can be used to resolve the contradiction. The mapping of resources is studied in Chapter 5.
3. Ideal Final Result: The Ideal Final Result is achieved when the contradiction is resolved. The desired features should be obtained without compromise. The use of the concept of ideality is considered in Chapters 6 and 8.

4. Functional analysis: Systems exist to create a functional output. Understanding systems functionally provides great insight into how the output is achieved and uncovers problems (contradictions), potential resources and, therefore, opportunities. The functional analysis methodology is covered in detail in Chapter 7.
5. Patterns of evolution: Systems evolve according to certain patterns, not accidentally. The patterns can be used many ways to get new ideas and predict the evolution of the system. Five important patterns are presented in detail in Chapter 10.
6. Innovative principles: These principles give concrete cues for solutions and illustrate what the patterns can mean. The list of 40 innovative principles is studied in Chapter 11.

Why introduce this model? The model for problem solving connects basic concepts and tools. The integrated system is more effective than the separate parts. It is based on the scientific research of TRIZ and on the feedback from many students of TRIZ over the past two decades.

References

1. Engeström, Y. 1987. *Learning by Expanding*. Helsinki: Orienta-Konsultit.
2. Shockley, W. 1976. The path to the conception of the junction transistor. *IEEE Transactions on Electron Devices*, ED-23, No. 7 (July), 597.

Chapter 3

Clarify the Tradeoff behind a Problem

3.1 Introduction

If a problem exists, clarify the tradeoff behind it. This is the first step in finding the real problem and good solutions.

We have already said that there are contradictions behind every difficult problem. The concept of a contradiction is very important. Participants in problem-solving training often wish for more help with contradiction analysis, because contradictions are at the core of the most challenging problems. In this chapter and the next, we will focus on the concept of contradictions.

Recall the model for problem solving (Figure 3.1). In this chapter, we analyze the conflict between two features. These are frequently called "tradeoffs" because the problem solver trades improvement of one feature against decline in another feature in the hope of finding a solution to the problem. In Chapter 4, we will study inherent contradictions, when one thing has two opposite properties. In traditional TRIZ books, tradeoffs are called "technical contradictions" or "engineering contradictions," and inherent contradictions are called "physical contradictions." We use lay terms because they make sense and it avoids having special definitions for TRIZ that do not always agree with everyday language.

- In this chapter, we discuss in detail why it is so useful to analyze tradeoffs. The benefit of TRIZ comes from solving difficult problems and that means resolving tradeoffs. To do this, it is necessary to formulate these tradeoffs or rewrite the problem in a form that makes the tradeoff obvious.
- We show how to formulate the tradeoff. The tradeoff appears in the system of the tool and the object. There may be different models of tradeoffs. The system has different features. When we have the problem, we have the tradeoff between features. A simple way to formulate the tradeoff between features is introduced.
- We consider complex problems that have many tradeoffs. The tradeoff can appear on different system levels and at different times. The selection of the tradeoff is discussed.
- The last part of the chapter presents five steps for problem clarification.

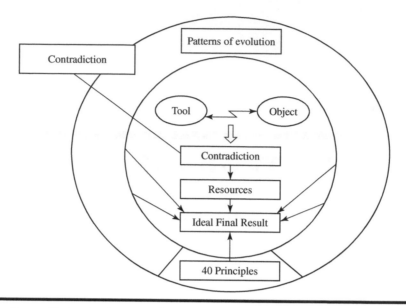

Figure 3.1   Contradiction is the core of a challenging problem.

3.2 What Are Tradeoffs and Inherent Contradictions?

In this and following chapters, we will often use the terms "tradeoff" and "inherent contradiction." These terms have the same meaning in TRIZ as they do in everyday English.

3.2.1 Tradeoff

When something good happens, something bad happens. Alternatively, when something good gets better, something else gets worse. Some examples of tradeoffs are:

- The product gets stronger (good) but the weight increases (bad).
- Software is made easier to use (good) but versatility decreases (bad).
- The hot coffee is enjoyable to drink (good) but can burn the customer (bad).
- Training gets more thorough (good) but requires more time (bad).
- The faster the automobile airbag deploys, the better it protects the occupant (good), but the more likely it is to injure or kill small or out-of-position people (bad).

3.2.2 Inherent Contradiction

One thing has two opposite properties: I want it cold, but I want it hot. I want it, but I don't want it. There are always inherent contradictions behind tradeoffs—sometimes they are obvious and sometimes they are hidden.

- The product should be thick (to get needed strength) and should be thin (to be light).
- Software should have very few options for ease of use and should have numerous options to be effective.

- Coffee should be hot for enjoyable drinking and should be cold to prevent burning the customer.
- Training should be lengthy to ensure good learning and should be very short to minimize demands on time.
- The automobile airbag should deploy quickly to save the driver or passenger, and should deploy slowly to minimize harm to small drivers or passengers.

As the examples show, the terms are easy to understand. If necessary, you can always go back to the examples to recall the definitions.

3.3 Why Analyze Tradeoffs?

Why are experts in engineering, business, and other fields so interested in contradictions today? The most obvious reason is that they need to solve problems. Moreover, the problems they need to solve have an important difference from the ones that people dealt with in earlier times.

Mankind has always resolved problems. Homer's *Odyssey* is, actually, a story of problem solving. First, Odysseus had to pass the Sirens and then sail between Scylla and Charybdis. The Sirens bewitched everybody, and there was "no homecoming for the man who draws near them. For with their high clear song the Sirens bewitch him, as they sit there in a meadow piled high with the moldering skeletons of men, whose withered skin still hangs upon their bones."[1] This problem was easy to solve because the goddess Circe gave good instructions: "to prevent any of your crew from hearing, soften some beeswax and plug their ears with it … if you wish to listen yourself, make them bind your hand and foot on board and place you upright by the housing of the mast, with the rope's ends lashed to the mast itself."[1]

The problem with Scylla and Charybdis was different. On the route rose two rocks. One was "the home of Scylla, the creature with the dreadful bark . … She has twelve feet, all dangling in the air and six long scrawny necks, each ending in a grisly head with triple rows of fangs, set thick and close and darkly menacing death. … No crew can boast that they ever sailed their ship past Scylla unscathed, for from every … vessel she snatches and carries off a man with each of her heads."[1]

On the other of the two rocks, "dread Charybdis sucks the dark waters down. Three times a day she spews them up and three times she swallows them down once more in her horrible way. Heaven keep you up from the spot when she does this because not even the Earthshaker could save you from destruction then."

Circe advised, "you must hug Scylla's rock and with all speed drive your ship through, since it is far better to lose six of your company than your whole crew."[1] Odysseus asked, "Could I not somehow steer clear of the deadly Charybdis, yet ward off Scylla when she attacks my crew?"[1] The goddess gave him a sound berating, called him an "obstinate fool," and continued, "Again you are spoiling for a fight and looking for trouble! Are you not prepared to give in to immortal gods?"[1]

Indeed, Odysseus had no defense against Scylla. The Sirens he passed without problems, following Circe's instructions, but to Scylla he lost six men: "Scylla snatched out of my ship the six strongest and ablest men. … I saw their arms and legs dangling high in the air above my head. … In all I have gone through as I explored the pathways of the seas, I have never had to witness a more pitiable sight than that."[1]

For thousands of years, people have accepted that there cannot be better ways to handle difficult problems. Had not the gods themselves warned humans against trying too much? Then, too, new solutions were not needed very often. Many problems were "Siren problems" that could be managed using what they already knew. However, this did not apply to all problems. During the same thousands of years, because of many cycles of trial and error, qualitative breakthroughs were accomplished: agriculture, clocks, the printing press, and other innovations. They clearly did not fit old thinking schemes. In Homer's terms, occasionally it was possible to sail without any losses past both Scylla and Charybdis.

Today there are more "Scylla-and-Charybdis" problems than ever before and requirements for the solutions are more stringent. It is no longer acceptable to lose six men, even if the rest of the crew will be saved. Likewise, it is not acceptable to solve a technical problem if the solution causes social problems or to solve a problem for your own customer but cause new problems for other people. Also, there is no time to wait for the trial-and-error method to eventually come up with a solution.

There are two types of problems:

1. Siren problems or problems that can be resolved straightforwardly using existing rules and instructions.
2. Scylla-and-Charybdis problems or those containing tradeoffs. That is, if you sail too close to Charybdis, you lose the whole ship. If you sail between them, you lose six people. If you sail too close to Scylla, you could lose more than six people. This is the original tough tradeoff.

TRIZ is a new approach that sees contradictions (tradeoffs and inherent contradictions) as sources of development. Resolving the conflicts in a system causes the development of the system. Resolving conflicts is the rationale behind successful inventions and innovations. If you want to move technology forward, you need to understand the conflicts. To do this, one method is to consciously clarify and intensify the conflicts or contradictions. Do not treat them as disorders that should be hidden; treat them as important clues to the solution.

3.4 Defining the Tradeoff

3.4.1 Tool and Object

The tradeoff arises in a system consisting of a tool and an object. All engineering systems (and many other systems that would not be considered engineering systems) are built from tools and objects. A knife, a saw, or a pair of scissors (or anything with a sharp edge) is a tool for working material. A car is a tool for carrying passengers and cargo. A transistor is a tool for switching electrical current on and off. A bee is a tool for collecting pollen.

Knowledge, models, and information are also tools. An advertisement is a tool to inform potential buyers. Computer-aided design (CAD) software is a tool for the design of products. A propane burner is a tool for heating objects. A committee or a team is a tool for making decisions in some companies. The model of TRIZ for problem solving is a tool for improving systems. We can describe the tool and the object using a simple diagram (Figure 3.2).

The tool is the component that is easiest to change when the problem is resolved. If the problem is how to decrease the wear of a saw that cuts metal, there are many possibilities—change the sharpness or the shape of the teeth, put coatings of tougher metal on the teeth, change the thickness of the blade, add a lubricant during cutting, and so on. The metal being cut (the object) most often cannot be changed or can be changed very little.

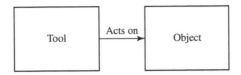

Figure 3.2   Tool, action, and object diagram that defines a system.

The object, on the other hand, gives constraints for the change of the tool. The material being cut limits the choice of cutting tool. Sometimes the object, too, can be changed in some way, for example, many objects can be combined—it may be more effective to cut a stack of glass plates than a single thin sheet of glass.

Action means that the tool does something that causes the object to change or to be maintained. It is fairly clear what is meant by changing the object (heating it, shaping it, etc.). However, what is meant by "maintaining" the object is not as straightforward. Actions sometimes maintain objects by keeping them in some state or condition that already exists. For example, a nail *holds* a picture by maintaining its position in 3D space. While there are other possibilities of maintaining an object outside of the function *holds*, the function *holds* is the most common of the maintenance actions:

- Bolt *holds* beam
- Clamp *holds* pipe
- Mortar *holds* brick
- Chair *holds* person
- Memory *holds* data

The statement of problems and solving those problems become much easier when the tools and the objects that are at the heart of the problems are isolated from the numerous other components of their system. For example, many gates are locked by a latch mechanism. The mechanism consists of a female part, a male part, and a pin. Female and male parts have a hole for the pin. Figure 3.3 shows how the mechanism works.

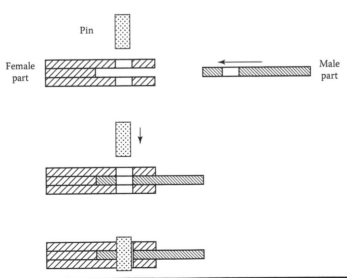

Figure 3.3   Tools and objects in the latching mechanism.

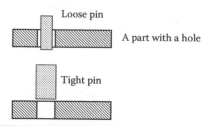

Figure 3.4   Limiting the view in the latching mechanism problem.

The male part is inserted into the female part. Then the pin is inserted into the hole. There is a small clearance between the pin and fixed parts, so that the gate can be easily locked and opened.

Latches of this kind work well on garden gates. However, when the same mechanism is used on a big ship, a problem appears. The vessel is constantly moving. Parts of the latch wear and may even fracture. In 1994, the ferry Estonia sank in the Baltic Sea and 852 persons lost their lives. The investigating commission found that the reason for the disaster was failure of a locking mechanism. As a result, a so-called visor was lost, water accumulated on the car deck, and the vessel capsized. (A "visor" on a ship is a type of gate that resembles the visor in a motorcyclist's helmet.)

If the pin is easy to lock and open (loose pin), the device gets less reliable. If the reliability is increased (by making the pin tight), opening and locking get more difficult.

To begin studying this problem, we will select the two components that disturb each other. Let us imagine a pin or peg and some part with a hole (Figure 3.4).

By limiting our view, the problem has become simpler and better defined. We see only two parts: a pin and a component with a hole. Gate, visor, actuators, electronic control devices, and numerous other components have vanished. Using common, everyday words instead of specialized technical terms can help with this step in the analysis.

A very simple solution to this problem was found and patented in the 1990s (yes, in the last decade of the twentieth, not the nineteenth century). The principle of the solution is shown in Figure 3.5. A conical pin replaces the cylindrical pin. The angle of the cone is selected so that the pin is tight enough but will not stick to the latch. (The proper angle was 15.4°—details depend on the thickness and the material of the latch and the pin.) Here are excerpts from the patent text: "The locking mechanism according to the present invention assists in eliminating problems arising from the deformation of large gates … invention includes an elongated guide slot part and an elongated locking part for securing within the guide slot part. The locking part and the guide slot part each include two long sides that are beveled in a similar manner for forming a tight fit. … Moreover, the locking mechanism reduces the amount of force needed to initially open a gate or hatch."[2]

The tradeoff is resolved. A conical pin is easy to insert and remove. In the locked position, it entirely fills the hole and will not wear.

One may ask: What is interesting in this solution? Would not every 10-year-old child find it? If this is your thought, try to answer the following two questions:

1. How was it possible that such a simple solution was not found until the 1990s?
2. What should be done to avoid big losses of time in seeking simple solutions that are badly needed?

In business problems, understanding the purpose of the activity may be even more important than in simple engineering tasks. For example, a company produces boilers, turbines, generators, and other energy technologies for industry. What is the object of the company's activity? In earlier decades, the answer would be simply to produce and sell boilers and turbines. Today, companies often prefer to say that solving the customers' problems is the real objective of their work. Indeed, the buyer of energy technology usually does not need equipment as such, but energy. Do the buyers of energy need just energy? Maybe they need solutions that help to save energy? This is why companies invest so much time and so many resources to develop mission statements—once the entire company has a common definition of its goal, it becomes much easier to judge any proposed action by whether it advances the goal.

Returning to the discussion of relatively simple systems, it is true (and sometimes confusing) that the same element of the system can be an object or a tool, depending on the problem. If the problem is how to get the ax to cut the wood chunk more effectively, the ax is the tool and the chunk of wood is the object. If the problem is that the edge of the ax is not sharp, the wood is the tool that dulls the blade and the ax blade is the object that is acted on. In some situations, we sell things to the customer, and in others, the customer provides data that modifies our actions. The customer can be the tool or the object.

For any particular situation, it takes some work to decide what is the tool and what is the object, but it will make your problem solving easier, so it is worth spending the time. It may be helpful to draw a picture or write the sentence "Tool acts on object," and then substitute the words from your problem.

3.4.2 Tradeoffs Everywhere

It is worth repeating that tradeoffs will inevitably arise in any system. The system consists of the tool and the object interacting with each other. In the top row of Figure 3.6, the straight arrow represents the situation without a tradeoff. Then the tradeoff appears (a flash arrow) and is solved (a big arrow shows a change in the system). The pattern repeats many times, as shown in the bottom row.

The tradeoff may appear clearly as a less-than-desirable action that accompanies a useful action. In the cutting tool, the edge works the material, but at the same time heats it. The material being worked wears the edge. The car transports people and cargo, but the noise of the car disturbs

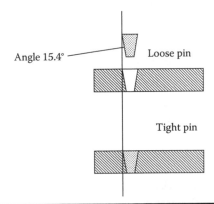

Figure 3.5   **The principle of the solution of the locking mechanism problem.**

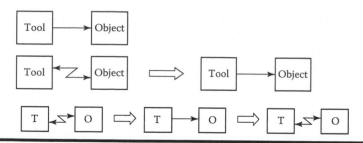

Figure 3.6 A system improves through the resolution of tradeoffs. In the top row, the straight arrow represents the situation without a tradeoff. Then the tradeoff appears (flash arrow) and is solved (big arrow shows a change in the system). The pattern repeats many times, as shown in the bottom row.

people and exhaust gases pollute the atmosphere. The transistor produces harmful heat while it does beneficial switching.

There are tradeoffs in the system even when there are no visible less-than-desirable effects. The electric car is silent and does not produce exhaust gases, but the power system occupies a great deal of the space in the car and reduces the load it can carry. No matter what kind of system is used to generate electricity (solar, conventional, or nuclear) to charge the batteries, equipment is needed to produce the energy, pollution is generated in the system, and additional machinery and energy are used to produce the infrastructure of the power system. Eventually, the batteries must be disposed of, which is a very big source of pollution. Each of these issues gives rise to tradeoffs.

To generalize, the system is bad simply because it exists. To have the capability of acting on the object (useful feature), the tool has dimensions and weight, consumes energy, and has other features that engender cost and harm. When the tradeoff is removed, you might think that the perfect system or the Ideal Final Result has been achieved. This achievement, however, is relative and temporary. New tradeoffs appear soon, either in the system itself or in other systems that are affected by it. Every new generation of transistors has occupied less space and consumed less energy than the previous one, but the higher density of transistors on the new chips requires more and more complex systems for connection and for heat dissipation. The electric car eliminates petrochemical pollution, but increases the problem of battery production and disposal. Only once a system has reached a state of complete ideality is there a possibility of having no contradictions within it.

3.4.3 Different Models of the Tradeoff

To describe the tradeoff between the tool and the object graphically, a variety of symbols are used (see Figure 3.7). The tool can interact in a useful way with one object and in a harmful way with another: a truck moves cargo, but at the same time wears out the road.

3.4.4 Features

Features and actions can describe tradeoffs. It is simplest to describe features. Most often, the development of a product means the improvement of the features. Usually, the user does not need the product itself, but needs the features. Consider the following examples.

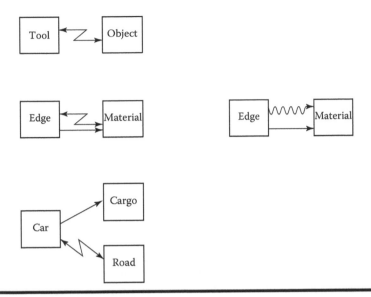

Figure 3.7   A variety of graphical models are used to show tradeoffs. The flash-like arrow represents the tradeoff in general. Sometimes, the flash represents negative action and the straight arrow a positive one. Edge cuts material, but also heats it. In some TRIZ books, wavelike lines often show a harmful action, particularly in the substance-field (Su-field) method. Additionally, in some software systems, blue arrows may show useful actions and red arrows may show harmful actions, so the presence of both colors in such systems would signify a tradeoff.

How can we improve a muffler in the conventional lawnmower powered by an internal combustion engine? List the features of the muffler:

- Dimensions
- Weight
- Noise absorption capacity
- Ease of manufacture
- Form, outer appearance

How can we thin out carrot seedlings in a small home garden? List the features of the thinning technology:

- Speed
- Precision
- Ergonomic level

We have spoken of the latch mechanism. How can we improve the pin? List the features of the pin:

- Dimensions
- Form
- Surface quality, manufacturing precision
- Ease of manufacture

- Ease of locking and opening
- Reliability

3.4.5 *The Tradeoff between Features*

The tradeoff between tool and object shows where the problem lies. It is useful to express the tradeoff as a conflict between two features. When the velocity of a car increases, safety worsens. A simple diagram, as shown in Figure 3.8, can illustrate the conflict between two useful features.

Often, the improvement of a useful feature is connected with the strengthening of a less-than-desirable or harmful feature (see Figure 3.9). When the velocity of a car increases, the consumption of fuel also increases.

We can use the features to describe the current state of the system and we can use them to describe the system that we want. If there is a conflict between the features or a tradeoff, we can use TRIZ to remove the tradeoff and create the new, improved system.

Frequently the analysis of functions (actions) can give valuable additional information about tradeoffs, if the functions can be expressed clearly. Often, however, one cannot find verbs and nouns that can satisfactorily describe the functions of even simple everyday systems. What is the function of a water tap? Is it to control water flow? "Control" is an extremely general verb that does not tell us how to detect the function of delivery of water. "To stop water flow" or "to regulate water flow" only partially describe the function. In contrast, one can easily list essential features of a water tap: water flow control, reliability, ease of use, decorative appearance, and others. Both features and functions will be used in the examples in the following chapters.

When people describe problems, often only the drawback is mentioned instead of the tradeoff. Insufficient safety or excess consumption of fuel is identified as the problem with a car and the tradeoff (fuel consumption increases when speed increases or safety decreases when speed increases) is not mentioned. Sometimes, desired features are formulated instead of the tradeoff. Safety and fuel economy are described as tradeoffs without specifying what feature is diminishing when safety or fuel economy improves. The problem solver tries to jump directly to the solution.

Analysis of both sides of the tradeoff is essential. It is not academic hairsplitting. When the connection between good and bad features is expressed, some features of the solution begin to appear. We know what tradeoff the solution should remove, although we do not yet know how.

In real situations, problems contain many possible pairs of tools and objects and many tradeoffs. We recommend trying the exercise in Table 3.1 to see the multiplicity of tradeoffs. It is suggested that you fill in the table with examples from your personal life and from your business

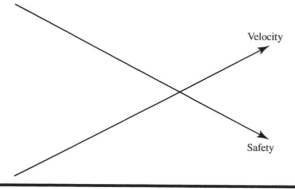

Figure 3.8   A conflict between two useful features.

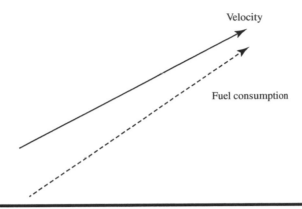

Figure 3.9   A conflict between useful and harmful features. Here, the dotted line means a harmful feature.

Table 3.1   Examples of Tradeoffs

When THIS Gets Better	THIS Gets Worse
The size of the warehouse increases	The accuracy of the inventory decreases
My family is happy with a vacation	It takes too much time from my work

experience. Which components and which tradeoffs should you select to get a good solution to your problem? In the following section, we will consider the selection of tradeoffs.

3.5 An Abundance of Tradeoffs

3.5.1 Where Does the Tradeoff Appear?

The problem statement is like the answers to a journalist's questions: who, what, where, when, why, and how. Answering the questions may reveal the tradeoffs. The level of the system is important, too. The system contains subsystems (e.g., the subsystems of a pair of sunglasses might be listed as the frame, the earpieces, and the lens) and itself might be a part of a higher-level system (e.g., pilot's flight gear/uniform) or macrosystem, also known as the "supersystem."

The muffler of the lawnmower is a system containing at least two subsystems: a casing and porous absorbing material. The muffler itself is a part of the lawnmower. The lawnmower, further, is a part of a park system or a garden system that contains grass, dirt, the gardener, and other systems (supersystem).

There are tradeoffs and solutions on all levels:

- The level of the muffler: The lower the level of noise, the thicker the layer of porous material. One possible way to decrease noise without thickening the muffler is using noise cancellation. An active noise-control device generates sound waves whose peaks correspond to the valleys of the undesired sound and vice versa. A system like this is used in some cars.
- The level of the lawnmower: This system contains casing, engine, exhaust tube, muffler, and other parts of the lawnmower. If noise is decreased using the muffler, the lawnmower becomes more complex. One obvious problem statement is how to change the casing so that it also does the job of the muffler.
- The level of a garden (or yard or park): The system contains at least the lawnmower and grass. The number of problems and solutions increases. The grass can be used in two different ways:
  - A Finnish inventor turned the exhaust tube down into the grass (see Figure 3.10). Grass worked as the sound-absorbing material. Noise decreased considerably. Hot gas also dried the grass so that it did not stick to the lawnmower.
  - When the appearance of the garden is improved using the lawnmower, the system gets more complex. If the grass does not need to be cut, we will not have to worry about the noise from the lawnmower. Grass that grows to a certain height and then stops has been developed and is being tested for commercial use. It can be said that the tool (lawnmower) has reached an ideal state by transferring its action (function) of cutting grass to the supersystem (the grass itself). In this way, there are no longer any contradictions within the tool (lawnmower) because it ceases to exist. Other examples demonstrate the same increase in ideality by way of transferring the action to the supersystem. The Japanese have hundreds of years of experience developing moss gardening. Moss carpet does not need any cutting. In trailer parks, people frequently paint rocks green and have no grass. Clearly, the best solution to the problem requires knowing what the use of the system will be.

It is useful to formulate problems and tradeoffs on more than one level (see Figures 3.11 and 3.12). The selection of the tradeoff depends on the constraints and conditions of the problem. If you need a quick and cheap solution, it may be best to begin with mechanical changes to the lawnmower or the muffler. If there are more resources available, solving problems on the micro- and

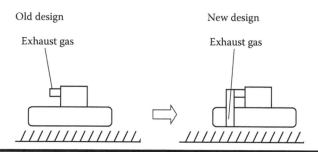

Figure 3.10 Grass working as sound-absorbing material.

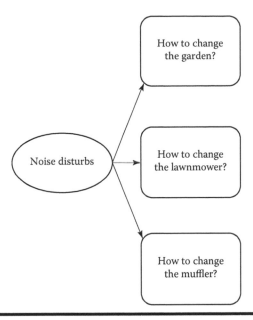

Figure 3.11   Formulate problems on more than one level.

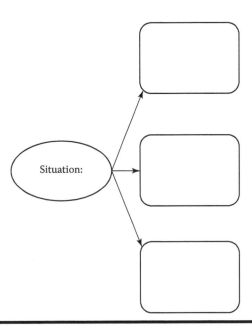

Figure 3.12   Exercise: Formulate problems on more than one level.

macrolevels (i.e., subsystem and supersystem) can be more exciting and can suggest changes that will influence the whole industry.

There is no way to guarantee a perfect problem statement. However, if you start with the simplest, most obvious statement, you will be able to improve it later. It is important to start.

3.5.2 When Does the Tradeoff Appear?

Problems can appear at different times in the life of the system. Another gardening example can illustrate this. Carrots are cultivated in a small home garden. The initial problem is how to make the thinning of carrot plants easier. Thinning is necessary so that there will be enough space between the plants as they mature to allow each plant to get enough nutrients and water. Before hurrying to solve this problem, it is useful to consider the stages of cultivation. For simplicity, we will look at only two stages:

- Seeding
- Thinning

Both stages contain their own problems and tradeoffs.

- Seeding: If every seed is planted precisely in the right place, no thinning is needed. If the gardener plants seeds very simply by hand, much time is needed. Time can be saved using seeding equipment, but then the drawback is the existence of the equipment, its cost, storage, maintenance, and such.
- Thinning: The same tradeoff as in seeding—thinning can be mechanized in some way and time saved, but the system becomes complex.

There is one beautiful solution that makes seeding simple and thinning unnecessary. Seeds are fixed on a biodegradable tape. The gardener places the tape in the furrow. The plants grow in exactly the right places (see Figure 3.13). You can, yourself, make a seed tape from paper towels, white glue, and small seeds (see http://lancaster.unl.edu/hort/youth/seedtape.shtml).

Sometimes a good solution is found in another stage of the process. That is why it is useful to formulate tradeoffs in different process stages. In this example, the problem of the thinning stage can be eliminated by a change in the seeding stage.

3.5.3 Nine Screens

The location of the problem in time and on the system level can be illustrated by a simple table containing nine screens (sometimes called "windows," "boxes," the "tic-tac-toe method," or the

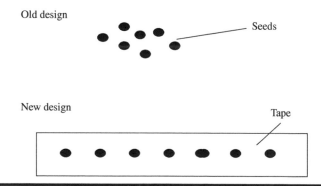

Figure 3.13   Seeds fixed on a biodegradable tape.

"system operator"). We show the examples of the lawnmower and the cultivation of carrots on the nine screens (see Figure 3.14).

The table in Figure 3.14 helps you to see the problems more clearly and sometimes suggests solutions very directly. In the carrot example, we considered present and past. How about the future? If the carrot row is not thinned, roots will be small. If we can find a use for "minicarrots," thinning is not needed. How about macrolevel? A tape with seeds can be considered a macrolevel system. How can this system be improved? For example, the tape could contain fertilizer.

Altshuller called the table "screens of talented thinking."[3] Usually, people see one screen: a system at present. A talented thinker sees at least nine screens: system, macrosystem, microsystem, and all three levels in past, present, and future.

It is important to note that some references to the nine screens will label the rows (top to bottom) as supersystem, system, and subsystem. Just as populations, groups, or gatherings do not refer to a specific and fixed set of people, the general terms of "supersystem," "system," and "subsystem" are equally nonspecific. A supersystem might be an aircraft operating by way of multiple subsystems, one of which is an engine. Further, the aircraft system operates with a supersystem called an airport (at least during part of the aircraft's operational process). Conversely, the fuel pump on the aircraft engine might initially be considered as the system that operates within a supersystem known as the engine. Further, one subsystem of the fuel pump could very well be defined as an electric motor. The definitions of system, supersystem, and subsystem for any analysis is dependent

	Past	Present	Future
Macrolevel		Lawnmower	
System		Muffler	
Microlevel		Porous substance	

	Past	Present	Future
Macrolevel			
System	Carrot: seeding	Carrot: thinning	Carrot: consumption
Microlevel			

Figure 3.14 **Examples of modeling by nine screens.**

upon the problem at hand and the benefits of analyzing the system at different hierarchical levels. Figure 3.15 presents a template for the use of the nine screens.

This is the simplest method for using the nine screens. The system of screens is an independent set of tools that can be used many different ways—to enhance your understanding of a problem and to help you expand the areas in which you can look for solutions. You can obtain many more ideas about using the nine screens from other sources.[4]

3.5.4 How to Decide Whether to Develop the System or Remove It

Sometimes, it may be simpler to remove the system than to develop it. In other words, sometimes a problem can be eliminated by simply eliminating the problematic components. In his historic work on brainstorming and idea generation, Alex Osborn suggested asking the questions: "What can we eliminate? ... Suppose we leave this out. ... Why not fewer parts?"[5] Usually, however, it is not possible simply to leave out a part or a stage of a process. The elimination of a part or operation means, in most cases, that some useful features and functions will disappear. By suggesting the removal of a part of a system or a process, we have created a tradeoff. If we can formulate the tradeoff explicitly, and then resolve the tradeoff so that the system will have fewer parts and operations and more useful features, we will have significantly improved the system. Conventional thinking often stops before considering this kind of tradeoff.

	Past	Present	Future
Macrolevel			
System			
Microlevel			

	Past	Present	Future
Macrolevel			
System			
Microlevel			

Figure 3.15   Exercise: Give examples from your personal life or your business situation using the nine screens.

What is the new perspective that TRIZ gives here? Instead of the suggestion: "Suppose we leave this out," we use new questions: "What are the good and bad features of this component or operation? What gets better and what gets worse, if we leave this out?" Some examples follow:

- What is good and what is bad about the spare tire in the car? It increases reliability, but at the same time occupies space. In some cars, the damaged tire is also the spare. It is designed so that, after a puncture, the car can be driven cautiously to the nearest repair shop.
- In the beginning of this book, we gave the example of the pipe with a T-joint. T-fittings are needed to make a complete pipe system (+), but they increase the number of parts (−). A collared pipe is the solution that provides the function of the T-joint without the complexity.
- Operations can also be removed. The harvester cuts and reaps at the same time; although, years ago, cutting and reaping were separate operations. Digital printing removes typesetting. On-the-job training removes classroom training.
- Storage operations and warehouses, once considered necessary, have been practically eliminated in many industries, such as electronics. Products are manufactured strictly on demand and then delivered directly to the point of use. However, there are also industries where storage problems persist. Much food is wasted in markets. Distributors keep extra food to ensure that buyers can find what they need, but some of the food will be left in the warehouse. How would you solve this problem?
- It was not long ago that anytime an item was purchased from a retail store a cashier would record the merchandise's cost, take payment for the merchandise, issue a receipt to the customer, and bag the merchandise. Now there are many situations where the buyer has the option to use an automated checkout service. The cashier is eliminated by having the customer scan the merchandise for pricing, pay, receive a receipt (all through an automated system), and then bag their own purchase. A very likely next step in the advancement of convenient checkout services is that the customer simply walks out of the store with the "purchased" item, where, by way of a radio frequency identification tag, the store scans the merchandise and directly transfers money from the customer's bank account through an interface with a payment application previously set up on the customer's mobile device. Receipts will be issued automatically in a similar over-the-Internet process. In this manner, the activities previously required to purchase merchandise are fully eliminated, completely replaced by electronic data transfer.

Removing the component or operation is not always possible. Often, some other solution for the tradeoff is needed, but this method, thinking about what would happen if we removed the component or operation, helps to find the real problem.

More detail on system trimming (component elimination) can be found in Chapter 7, Section 7.6.2.

3.5.5 How to Identify the Right Problems to Solve

An old proverb reads: "It is more important to do the right things, than to only do things right." Unfortunately, repeating this and similar statements helps very little. Problem selection in the past has often been the expert's subjective choice.

TRIZ is revolutionizing the technology of problem solving, and it is also changing the methodology of developing the problem statement. Instead of subjective or arbitrary formulations, using TRIZ, we have a precise definition. A real problem contains tradeoffs.

The analysis of tradeoffs is only one of the tools in the toolkit that TRIZ provides for developing a precise problem statement. All the tools presented in this book can be used for both problem finding and problem solving. You can apply the tools one at a time as you read each chapter, but you will enhance your problem-analysis and problem-solving skills by using all the tools together.

In this section, we focus on tradeoffs. Finding the best problem to work on means selecting a relevant conflict from many tradeoffs.

First, consider the available time and resources. In principle, there are many alternatives. In practice, the choice is much more limited. In the lawnmower example, the most dramatic improvement is to get nice-looking grass without a lawnmower. This might be the best solution, but it requires a large investment of research money over a considerable period, and then another large investment to introduce the new ground cover to the market. If the lawnmower company needs a quick or a cheap improvement, it might be better to examine the muffler and the elements of the system that interact or could interact with the muffler. The same can be said of making choices for improvements at different stages of a process. Sometimes, one can solve the problem by going to the past (see the example of carrot cultivation) or to the future, and sometimes not.

Second, the problem appears between certain components. That simplifies the choice of the problem. You simply select the components touched by problems. The pin and the part with a hole disturb each other in the latch problem. That is why they are selected. A gate and actuators, for example, are omitted. They are not connected with harmful interactions or features.

These recommendations, however, do not guarantee the right choice for the problem. If we can find a ground cover that does not need cutting, the right task is to remove the lawnmower from the system. If we cannot, the right task is to improve the lawnmower. That is, to formulate the right problem, we should know the solution. To get the solution, it is very important to state the right problem. Behind the popular phrase "a problem well stated is half-solved" lies a silent assumption that one can state the problem well. This assumption is wrong. No expert in the world can state the problem exactly at first. Is this a deadlock? What, then, should the TRIZ practitioner do?

There is a well-proven, practical approach. State the problem that seems to be most reasonable based on existing knowledge and try to solve it. During the attempt to get the solution, new obstacles, as well as new opportunities, will appear. This new knowledge allows us to modify the problem statement. This kind of generation of new ideas may remind us of the exploration of unknown continents in the past or of space research today. We do not know what is out there, but we have some basis for the assumption that it is worth going. Later explorers (or the first explorers on later trips) will have better ideas about what they should look for, so they can bring better tools or people with more specialized skills (see Figure 3.16).

3.6 From the Problem to the Tradeoff

Let us summarize the points in this chapter. There are five steps for the clarification of a problem:

1. Describe pairs of tools and objects and the action that links them. Select one pair. Explain why you picked this tool and object.
2. Describe features and conflicts between them.
3. Select one tradeoff.
4. Explain why you identified this tradeoff.
5. Describe the tradeoff graphically and in words.

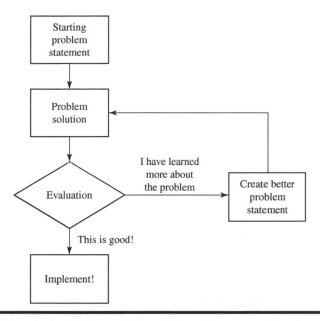

Figure 3.16   Flowchart for getting started in problem solving. In a flowchart, a box is an action and a diamond is a decision. The arrows show the order in which things are done. This flowchart describes the method of starting with a preliminary problem statement and then developing better problem statements as you learn more about the problem.

The following tables and figures contain the summaries of some examples: noise problem in a lawnmower (Table 3.2 and Figure 3.17), cultivation of carrots (Table 3.3 and Figure 3.18), and problems in the latching mechanism (Table 3.4 and Figure 3.19). Table 3.5 is a template for the study of your own problems.

3.7 Standardization of Tradeoff Statements

Standardization is useful in many technologies and the same is true in TRIZ. A useful standardization for writing tradeoff contradiction statements is as follows:

1. Use the words IF, THEN, and BUT in the statement.
2. IF describes a situation or relationship about the problem being analyzed.
3. THEN states what is good about the situation or relationship.
4. BUT states what is bad about the situation or relationship.

Example standardized tradeoff statement are as follows:

- IF a lawnmower uses a muffler for noise suppression
- THEN noise levels are reduced (good)
- BUT backpressure on the engine is increased (bad)

- IF carrot seeds are sewn by hand
- THEN the work "tools" are simple (good)
- BUT cultivation requirements are increased (bad)

Table 3.2   Constructing the Model of Tradeoffs Using the Lawnmower Example

1.	Describe pairs of tools and objects.
	a. Muffler and noise (or air that vibrates)
	b. Noise and a person (disturbed by noise)
	c. Engine and exhaust gas (engine moves gas)
2.	Select one pair. Explain why just this tool and object are selected.
	We select the pair "muffler and noise" because we want to select a limited problem. We want to remove the harmful feature with minimal changes to the system.
3.	Describe features of the selected system of a tool and an object. Describe conflicts in this system.
	Features: Dimensions, weight, absorption capacity, ease of manufacture, form, and so on
	Conflicts between features:
	a. When noise absorption improves (+), the dimensions of a muffler increase (−)
	b. When noise is suppressed (+), the number of parts increases (muffler needed) (−)
4.	Select one pair of conflicting features. Explain why just this tradeoff is selected.
	We select the formulation: "When noise is suppressed (+), the number of parts increases (muffler needed) (−)."
	The reason: A limited problem, very big changes not needed, at the same time the goal is more than simple optimization of the size.
5.	Describe the tradeoff graphically and in words.
	Graphically: see Figure 3.17.
	In words: When noise absorption capacity improves, dimensions (of the muffler) and number of parts (the muffler is needed) are increased, too. If the system is simplified (or muffler made smaller or removed totally), the noise absorption capacity is lessened or lost.

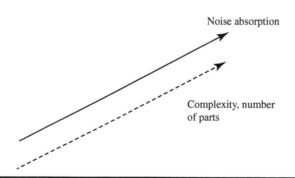

Figure 3.17   Tradeoff in the noise suppression problem.

Table 3.3   Constructing the Model of Tradeoffs—How to Cultivate Carrots

1.	Describe pairs of tools and objects.
	a. Hand and seed
	b. Hand and carrot
	c. Seeder and seed
2.	Select one pair. Explain why just this tool and object are selected.
	Hand and seed. The constraints of the problem allow us to make changes easily in the early stages of the process. The simplest system is selected.
3.	Describe features of the selected system of a tool and an object. Describe conflicts in this system.
	Features: Accuracy, speed, convenience
	Conflicts between features:
	a. The more accurately the carrot is seeded (by hand), the lower the speed
	b. The higher the speed when seeding precisely (by machine), the more complex the equipment
4.	Select one pair of conflicting features. Explain why just this tradeoff is selected.
	We select the first formulation: "The more accurately carrot is seeded (by hand), the lower the speed."
	The reason: Try the simplest case first.
5.	Describe the tradeoff graphically and in words.
	Graphically: see Figure 3.18.
	The more accurately the carrot is seeded (by hand), the lower the speed.

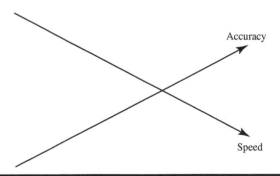

Figure 3.18   Tradeoff in the carrot cultivation problem.

Table 3.4   Constructing the Model of Tradeoffs—How to Improve the Latching Mechanism

1.	Describe pairs of tools and objects.
	a. Female and male parts
	b. Pin and a part with a hole
2.	Select one pair. Explain why just this tool and object are selected.
	Pin and a part with a hole. The problem, wearing, occurs between a pin and a part.
3.	Describe features of the selected system of a tool and an object. Describe conflicts in this system.
	Features: Dimensions, form, surface properties, manufacturing precision, ease of use, reliability
	Conflicts:
	a. If the pin is made easy to lock and open (loose pin), the device gets less reliable. If the pin is made tight, opening and locking get difficult
	b. If the pin is machined very precisely so that it is both more reliable and easy to use, manufacturing gets complex
4.	Select one pair of conflicting features. Explain why just this tradeoff is selected.
	We select the first formulation: "If the pin is made easy to lock and open (loose pin), the device gets less reliable. If the pin is made tight, opening and locking get difficult."
	The reason: The system that does not need precise machining is simple.
5.	Describe the tradeoff graphically and in words.
	Graphically: see Figure 3.19.
	If the pin is made easy to lock and open (loose pin), the device gets less reliable or wearing gets worse.

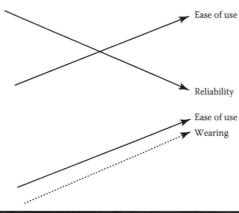

Figure 3.19   Tradeoffs in the problem of the latching mechanism. Two pictures illustrate that the same conflict can be visualized in different ways.

Table 3.5 Construct the Model of Tradeoffs—Consider Your Own Problem Situation

1.	Describe pairs of tools and objects.
2.	Select one pair. Explain why just this tool and object are selected.
3.	Describe features of the selected system of a tool and an object. Describe conflicts in this system.
4.	Select one pair of conflicting features. Explain why just this tradeoff is selected.
5.	Describe the tradeoff graphically and in words.

- IF the fit between a pin and slot is tight on a latching mechanism
- THEN the pin will not loosen easily (good)
- BUT it will be difficult to remove the pin when needed (bad)

- IF a kitchen knife is very sharp
- THEN it can easily cut food items (good)
- BUT severe injuries to fingers can occur (bad)

- IF the fuel tank is large
- THEN the vehicle has a long range (good)
- BUT the size and weight of the vehicle is increased (bad)

- IF a large quantity of cashiers is used
- THEN customers have a short checkout time (good)
- BUT the business must have additional employees on shift (bad)

- IF poison is used to kill weeds
- THEN the weeds are cleared effectively (good)
- BUT unwanted damage will occur to wildlife (bad)

- IF a training program is long and detailed
- THEN the student can be well trained (good)
- BUT the student has less time to support other activities (bad)

- IF the screen on a smart phone is large
- THEN a lot of information can be displayed (good)
- BUT the phone's size is also large (bad)

3.8 Summary

Why study contradictions (tradeoffs and inherent)? The two kinds of problems are Siren problems and Scylla-and-Charybdis problems. You do not know which kind of solution to use if you do not know which kind of problem you have. Scylla problems contain contradictions. Resolving contradictions in a system causes the development of the system.

To find the tradeoff, first select the tool and the object. Then, describe the important features of the selected system of the tool and the object. Select the pair of conflicting features that best characterize the problem. Use different verbal and visual models for describing contradiction.

There are many tradeoffs. Ask where and when the contradiction appears. The scheme of nine screens or windows helps to answer the questions when and where. Decide whether to develop the system, or remove it by formulating the contradictions that appear if a component or operation is removed. How can you find the right problems to solve? Try a variety of models and formulations of tradeoffs first to state the problem that is most reasonable and then modify the problem statement, if necessary.

Go from the vague problem situation to the clearly formulated tradeoff: describe pairs of tools and objects and the action that links them. Select one pair. Describe features and conflicts between them. Select one tradeoff. Describe the tradeoff graphically and in words. Sometimes, the formulation of the tradeoff gives you the idea for the solution. Sometimes, you need other tools, which will be presented in later chapters of the book. Even if you have a solution that you like, as a result of this method of clarifying the tradeoff, it would be a good idea to use the tools presented in later chapters to develop more solutions. One easy and effective set of tools are the 40 Principles. Additionally, to find the correct principles, there is another tool referred to as "Altshuller's matrix." The matrix allows the analysis of typical contradictions and shows standard solutions (selected 40 Principles) to them. Both the 40 Principles and the matrix will be presented in detail in Chapter 11.

Use the standardized form (If, Then, But) when writing tradeoff contradiction statements where:

- IF describes a situation or relationship about the problem being analyzed.
- THEN states what is good about the situation or relationship.
- BUT states what is bad about the situation or relationship.

References

1. Homer. 1991. *The Odyssey*, trans. Rieu, E. V. London: Penguin Books, 180–186. All citations to *The Odyssey* are to this volume.
2. Lahtinen, M. U. P. and Holtta, P. J. 1999. US Patent 5,875,658
3. Altshuller, G. S. 1984. *Creativity as an Exact Science*. New York: Gordon and Breach Science Publishers, 120.
4. Mann, D. L. 2009. *Hands-On Systematic Innovation*, 2nd Edition. Clevedon, England: IFR Press
5. Osborn, A. F. 1963. *Applied Imagination*. New York: Charles Scribner's Sons, 267.

Chapter 4

Moving from Tradeoff to Inherent Contradiction

4.1 Introduction

The more extreme the conflict you imagine, the better the solution you get. An example of an extreme conflict is the following: there should be many languages in the world to get cultural diversity. But, there should be only one language to make communication easy. In previous chapters, we briefly defined two different types of contradictions: tradeoff and inherent. We have considered tradeoffs in detail. In this chapter, we learn how to move from the tradeoff to the inherent contradiction, and how to intensify the contradiction.

Why, in addition to tradeoffs, do we need another way to express contradictions? The formulation of inherent contradiction is beneficial for several reasons:

1. The formulation of the inherent contradiction captures a higher level of abstraction than the related tradeoff contradiction does. This higher level of abstraction produces better focus on the problem to be solved and results in more elegant (ideal) solutions. There are always many problems and tradeoffs, but not all problems are equally important. One problem is key. The solution to this problem leads to the solution of others. Recall the fast-food example in Chapter 1. Old drive-in restaurants had many problems and tradeoffs: slow service, high costs, uneven quality, not always the best reputation, and others. The whole bundle of drawbacks resulted from one inherent contradiction: the restaurant needs many service people to carry out the operations (prepare and serve food, clean the restaurant, etc.), but the restaurant needs very few people to make it easy to coordinate all activities.

2. The formulation of the inherent contradiction includes one element of the good solution. The inherent contradiction should be removed. The self-service concept clearly removed the contradiction "many–few service people." There are no service people. At the same time, we can say that there is one service person to each customer—himself or herself.

3. If we can focus on one key problem, we can present problems and their solutions much better to all possible customers, as well as management, colleagues, and partners.

In this chapter, you will first learn how to formulate inherent contradictions. Instead of the two conflicting features that we worked with when we used the tradeoffs, we will have only one feature with incompatible values (big–small, many–few, long–short, much–little). Next, you will learn how to further intensify the contradiction. Often, the contradiction can be made sharper, or more extreme, which will help point the way to better solutions. Finally, this chapter will present examples of how to go from the visible drawback to the intensified inherent contradiction.

4.2 How to Formulate the Inherent Contradiction

So far, we have described contradictions between a tool and an object or, generally, in the system of the tool and the object. However, real problems frequently are not simple, and there are conflicts within each object, as well as between them (see Figure 4.1).

To act on the object, the tool should have certain weight, dimensions, energy consumption, and other features. At the same time, we want to minimize weight, size, and energy losses. The tool should have different and incompatible values of a single parameter. Examples of typical conflicts:

- Big–small: To be comfortable, the house should be big, and to decrease the consumption of material and energy, the same house should be small.
- Heavy–light: Strength, reliability, and safety require that a vehicle should be heavy, but energy economy requires that the same vehicle should be light.
- High energy consumption–low energy consumption: Machinery and tools need to use energy to work on objects, and they should use as little energy as possible.
- Generally, to improve feature A, an element should have a certain property, but to improve feature B, the same element should have the opposite property.

Inherent contradiction formulation:

Want *some element* to be in some *condition* because …
    And
Want *the same element* to be in *the opposite condition* because …
  As an example:
Want the house to be big because it will be comfortable
    And
Want the house to be small in order to consume less materials and energy

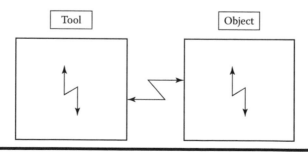

Figure 4.1 Conflicts within and between components. Tradeoffs (between components) may be difficult to solve directly because the components should have two opposite properties.

Just the formulation of the inherent contradiction, incompatible values of the same parameter, can guide us to ideas for solutions, as we can see in our examples. In the lawnmower noise problem, we had a tradeoff: if noise absorption improves, the size of the muffler increases. We can easily formulate the inherent contradiction "big muffler–small muffler."

Want the muffler to be large to improve noise absorption
　　And
Want the muffler to be small to reduce the size of the lawnmower

In the example of carrot cultivation, we have a tradeoff: the more precisely we seed carrots (by hand), the more time is needed (the slower the speed). Here, a little bit more thinking may be required to find the inherent contradiction. If we plant many seeds, we have a tradeoff between precision and speed. If we plant only two or three seeds, this tradeoff disappears, but we will have another one: we can seed precisely and not much time is needed, but the crop will be small. So, we have the inherent contradiction "many seeds–few seeds."

Want to plant a lot of seeds to increase the size of the crop
　　And
Want to plant a few seeds to increase the precision of the planting

In the seeding problem, the conflict is connected with the object, not with the tool. Objects often contain contradictions such as "many objects–few objects." There can also be an inherent conflict with the tool (the person planting the seeds). The planter wants high speed for a large crop, but low speed for precision planting, so the work of thinning the carrots later can be avoided.

In the example of the pin in the latching mechanism, we had a tradeoff. If the pin is made easy to lock and open, wear gets worse. The inherent contradiction is "loose pin–tight pin," or "big clearance–small clearance" (between pin and part).

Want a tight-fitting latch to reduce wear caused by movement
　　And
Want a loose-fitting latch to make it easy to open and close

Conflicts do not have to deal with physical objects. In the example of customer service, we need training so that employees can offer good service. However, while employees are attending the training class, they are not doing the work that their company and their customers need done. An inherent contradiction behind this tradeoff is "long training–short training." At the extreme, this becomes "continuous training–no training."

Want long training periods to improve the employee's performance
　　And
Want short training periods to increase the employee's work time

Using water to fight fires is a technical example that is familiar to everyone. However, the more water used, the more equipment needed. If the fire is put out, large amounts of water can damage the parts of the building that were not damaged by the fire. In some cases, the water does more damage than the fire. The inherent contradiction is "much water–little water."

> Want a lot of water to extinguish the fire
>> And
> Want a little water to not cause additional damage

Not long ago, a simple and interesting solution was developed. Water is atomized and sprayed in very small microdrops, or mist. Much less water is needed to extinguish the fire, and there is much less damage caused by the water itself. Of course, there is the new contradiction in the complexity of the atomizing equipment. That contradiction is partially eliminated by virtue of decreased water flow and high pressure—the size of tubes and other equipment can be decreased. The same volume of water gives 8000 times more drops than a conventional sprinkler and 200 times more than low-pressure mist equipment (Figure 4.2). Because the extinguishing efficiency depends mainly on the number of drops, pipe sizes can be radically decreased. For more information, see http://www.marioff.com/.

4.2.1 Present and Absent

One important group of inherent contradictions is made up of situations in which the object should be present to get some useful feature, and absent to keep the system simple or to avoid some other problem. In Section 3.5.4, we presented examples of tradeoffs that arise if we want to remove parts or eliminate operations from a system. Let us reexamine some of these tradeoffs and look at the inherent contradiction behind each one.

- Scenario One:
  - Tradeoff: A spare tire in the car increases reliability (+) but occupies space (−).
  - Inherent contradiction: A spare tire should be present in order to avoid being stranded *and* absent in order to create more cargo space. One beautiful solution is to use the normal tire as the spare by making it able to survive a blowout (inherent contradiction solved), rather than making the separate spare tire smaller, as in some cars (compromise).

- Scenario Two:
  - Tradeoff: If the complexity of a plumbing system is improved and liquid is distributed to more points (+), a greater number of T-joints are needed (−).
  - Inherent contradiction: A T-fitting should be present in order to expand the system, *and* the same fitting should be absent in order to increase the number of parts in the system.

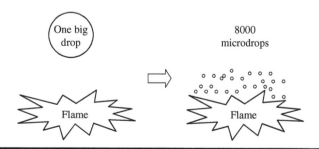

Figure 4.2 The conflict "much water–little water" resolved.

Table 4.1   List Your Ideas about How to Resolve Contradictions

Problem	Ideas
Tire reliability	
Plumbing parts	
Extra food	

■ Scenario Three:
  – Tradeoff: If a plentiful supply and good assortment of food is always available in the warehouse (+), considerable food will be wasted (–).
  – Inherent contradiction: Extra food should be present so that it is always available, *and* extra food should be absent to insure it never goes bad.

Do you have some ideas about how to resolve these contradictions? Write them down in Table 4.1 before you read the rest of this paragraph. Do not look at the following suggestions until you have written your own. There are many ways to solve each of these problems, and which idea is best will depend on the circumstances.

■ How can we eliminate the spare tire? Some truck companies do not have any spares in the vehicle. When a blowout happens, the driver calls the service center and gets a new tire. It is cheaper to wait a little for the new tire than to carry an expensive extra tire all the time.
■ How can a T-fitting be present and absent? A collar attached to the tube can function as a T-fitting.
■ How can extra food be present and absent? Maybe you can find buyers who will buy extra food near the last day it can be sold at discount prices. The seller could recoup some of the cost of the food and prevent waste. Would some food-on-demand concept work? In that scenario, all food could be sold at a normal price. Maybe extra food could be presented to a children's hospital or to a charity. In this case, it stops being extra food and becomes socially useful, like the Second Harvest program in many US cities.

Tradeoffs and inherent contradictions appear in all endeavors—engineering, commercial, organizational, educational, social situations, and such.

In the 1950s, Altshuller wrote that finding and resolving contradictions is essential in problem solving. During the past few decades, the same contradictions have been discovered in science, engineering, and business situations that, on the surface, appeared to be very different from each other.

Peters and Waterman wrote the bestseller *In Search of Excellence* in 1982. At least one chapter in the book is absolutely true today—perhaps more true today than it was 35 years ago. In Chapter 4—Managing Ambiguity and Paradox, they write: "Most important, we think the excellent companies, if they know any one thing, know how to manage paradox."[1]

Table 4.2 Adding Examples of Inherent Contradiction

Study examples of the inherent contradiction. Add three more examples of your own. Try one each from your business life, your personal life, and your community.

Let us go from business to dramaturgy. Syd Field writes in *The Screenwriter's Workbook*: "Drama is conflict; without conflict, there is no action, without action, no character, without character, no story, without story, no screenplay."[2]

It is useful to collect your own examples of inherent conflicts (see Table 4.2).

4.3 How to Intensify the Inherent Contradiction

We can sometimes make the internal conflict we just formulated much stronger. This will make the problem seem unsolvable—you might think this is a bizarre technique, but it can lead to great solutions. Think of solutions such as:

- A spotlight is present where it is aimed (illuminates) and is absent where it is not aimed (does not illuminate).
- The transistor can be a conductor and an insulator at different times.
- The food strainer is present for the food (stops the food) and absent for the water (allows passage).

Let us return to examples studied earlier. The contradiction can be intensified:

- Instead of the small muffler we require, try the idea of no muffler at all.
- Instead of the few seeds we require, try the idea of a single seed.
- Instead of the small clearance we require, try zero clearance.
- Instead of the short training we require, try no training time at all.
- Instead of decreasing the amount of water to fight the fire, try no water at all.

Intensifying the contradiction is the key to the solution. When the grass is used to absorb noise, we have either no muffler or a very big muffler. When many seeds are fixed on tape, we have one seed planted at a time. A conical form pin has no clearance during work and sufficient clearance when one needs to close or open the mechanism. Training embedded in work may be long lasting, although classroom training may not even take place. Water mist has many properties of a gas. It is practically dry, that is, it does not cause water damage, yet it is wet enough to suppress fire. There is a qualitative difference between little water and no water.

These examples show how the TRIZ technique of forcing the contradiction to the extreme can be a technique for breakthrough. Analyzing the extreme contradiction helps one break out of conventional thinking into the realm of ideas that get rid of the contradiction and solve the problem.

Table 4.3   Examples of Intensified Conflict

Study examples of intensified conflict. Add three more examples of your own. Try one each from your business life, your personal life, and your community. You can intensify, if possible, conflicts you have formulated in Table 4.2 or add totally new examples.

If the wildness of extreme contradictions makes you uncomfortable, remember that they are only needed when new ideas and solutions are needed. They are supposed to shake you up, to get you to examine possibilities outside the system that caused the problem in the first place. If you have an existing solution to the problem that needs minor improvement, go ahead and improve it—do not go to the extreme. To get comfortable with extreme conflicts, try the exercise for intensifying conflict in Table 4.3.

For further study of inherent contradictions, you can find material in other books. Altshuller's *And Suddenly the Inventor Appeared*[3] contains a short introduction to the concept of the physical contradiction. The same topic is considered in Savransky's book[4] beneath the term "physical point contradiction."

4.4 Summary

Let us review how we applied the idea of the hidden inherent contradiction and the intensified contradiction to our examples (Tables 4.4 through 4.6).

Continue the study of your own problems. Summarize the intensification of the contradictions. Does the intensification give you any new ideas about solutions? See Table 4.7.

Altshuller published the first article on TRIZ in 1956 with his friend and colleague Shapiro. (The paper was not published in English until 2000.) Defining the critical contradiction and determining the immediate cause or contradiction were named as essential stages of problem solving.[5] Later, in the 1970s, Altshuller expanded the definition from one type of contradiction to two: tradeoff and inherent (technical and physical contradictions by the old terminology). If you are learning TRIZ now, you can start with these important concepts and save 20 years.

Numerous problems and tradeoffs can be boiled down to a single inherent contradiction. Inherent contradictions represent higher-level abstractions of problems and, therefore, usually result in better and more elegant solutions. Inherent contradictions are formulated as the following:

Want condition A for element X because _____
    And
Want condition opposite of A for element X because _____

Table 4.4   Reducing Lawnmower Noise—Intensifying Contradiction

Modeling Steps	Example: Reduce Lawnmower Noise
Visible drawback	The lawnmower is too noisy
Tradeoff: The conflict between two features	When noise absorption improves, the size of muffler and number of parts increase
Inherent contradiction	Big muffler–small muffler
Intensified inherent contradiction (if it can be intensified)	Big muffler–no muffler

Table 4.5   Cultivating Carrots—Intensifying Contradiction

Modeling Steps	Example: Make It Easy to Grow Carrots
Visible drawback	Thinning of carrots is an arduous job
Tradeoff: The conflict between two features	The more precisely carrot seeds are planted, the slower the speed
Inherent contradiction	Many seeds–few seeds
Intensified inherent contradiction (if it can be intensified)	Very many seeds–one seed

Table 4.6   Improving Latching Mechanism—Intensifying Contradiction

Modeling Steps	Example: Improve Pin-Type Latch
Visible drawback	Latching mechanism wears and can even fracture
Tradeoff: The conflict between two features	If the pin is made easy to lock and open, wear gets worse
Inherent contradiction	The clearance between the pin and the part should be small, and the clearance should be big
Intensified inherent contradiction (if it can be intensified)	There should be no clearance between the pin and the part, and there should be big clearance

Table 4.7   A Template for the Study of Your Own Problems

Modeling Steps	Your Example
Visible drawback	
Tradeoff: The conflict between two features	
Inherent contradiction	
Intensified inherent contradiction (if it can be intensified)	

It is advisable to intensify the inherent contradiction as much as possible. The more extreme (intensified) the contradiction, the better the solution. The solution of the inherent contradiction removes many problems and gets many benefits in one stroke.

Terms such as "contradictions," "paradoxes," and "conflicts" are now increasingly used across industries and in many business areas.

References

1. Peters, T. J. and Waterman, R. H. 1982. *In Search of Excellence*. New York: Harper & Row, 91.
2. Field, S. 1984. *The Screenwriter's Workbook*. New York: Dell, 31.
3. Altshuller, G. S. 1996. *And Suddenly the Inventor Appeared*. Worcester, MA: Technical Innovation Center, 21.
4. Savransky, S. D. 2000. *Engineering of Creativity*. Boca Raton, FL: CRC Press, 235.
5. Altshuller, G. S. and Shapiro, R. B. 2000. Psychology of inventive creativity. *Izobretenie*, II, 23–27.

Chapter 5

Mapping Invisible Resources

5.1 Introduction

In Chapter 2, we defined resources as things that are available but are not being used, or not being used fully—sometimes, we cannot "see" them because we have too many biases. In Andersen's fairytale, a little child saw that the emperor had no clothes. The ancient Mayans used wheels for toys and obviously knew how to make wheeled vehicles, but they never built them for any other uses. We recommend you review the examples in Section 1.2 before proceeding with this chapter.

Our use of the word "resource" in this book is nearly the same as in common language. Our list of resources includes materials and energy, human resources (HR), information resources, and so on. The word "nearly" indicates an important limitation: we are interested in idle resources available in the system and its environment, not in all resources. We are interested in free or very cheap resources, not in expensive additions.

First, we consider the invisible resources of systems. All systems have gray zones or proximal zones of development. They are areas where we can find solutions that have the potential to be developed, but have not yet been developed. They are zones where business opportunities lie.

Second, we look at the benefits from resource analysis. Understanding resources will help you in many ways. The analysis of resources in a situation can independently stimulate new ideas. Understanding resources can resolve the inherent contradiction that creates a problem in the first place. The analysis of resources will help you to foresee the evolution of the system and to understand customer needs that you have not previously identified.

Third, we present a simple, handy classification of the most useful resources. The tool, object, environment, and macro- and microlevel systems are resources classified by their relationship to the system. On each of the system levels, a variety of resources may be available: substances, energy, space, time, and such. Other resources, such as information, people's skills, and solutions, used by other industries are also available.

Fourth, we study the seven most important resource groups in detail.

The final part outlines using resource analysis for explaining undesirable phenomena, for example, occasional faults in products with no visible reason.

In the model of problem solving (see Figure 5.1), resources work as a bridge between the contradiction and the Ideal Final Result. As we have already said, resource mapping can also be used as an independent tool.

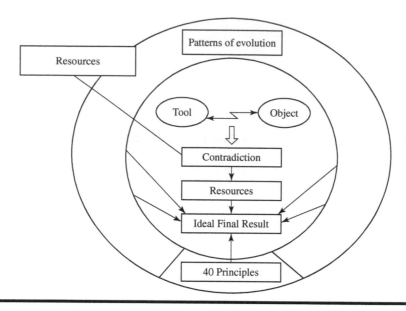

Figure 5.1   Consider all the resources. The mapping of resources can be used both as a step after defining the contradiction and as an independent tool.

5.2 Invisible Resources

Frequently, everything needed to resolve the contradiction is available, but the conflict has not yet been solved. Resources lie in a zone between the current level of technology and a more ideal, but feasible, level. The relationships among contradiction, resources, and the Ideal Final Result are shown in Figure 5.2.

An analogy from a person's development may help explain the use of resources. People cannot learn everything, but they are capable of learning some skills in given time. For example, an average person can learn a foreign language over a period of time. The skills that people do not yet have but can develop for themselves are called "the zone of proximal development." Correspondingly, we can say that resources define the zone of proximal development in technology—solutions that can be developed but have not yet been developed.

In the evolution of the lawnmower, we can easily see the three concepts we have spoken of: contradictions, resources, and Ideal Final Result.

1. Lawnmower contradiction
   a. Tradeoff Contradiction:
      IF a lawnmower uses a muffler for noise suppression
      THEN noise levels are reduced (good)
      BUT backpressure on the engine is increased (bad)
   b. Inherent Contradiction:
      Want a huge muffler to eliminate the engine noise
      And
      Want no muffler to increase the efficiency of the engine
2. Resources: Grass and a small duct directing airflow are available, but are unused resources.
3. Ideal Final Result: Grass is used as a muffler.

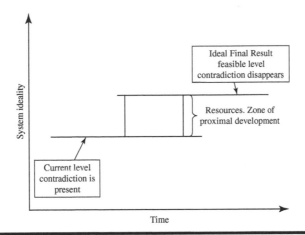

Figure 5.2   The relationships among contradictions, resources, and the Ideal Final Result. In time, new resources appear. After a time lag that, sometimes, can be very long, they are found and used. One benefit from resource mapping is reducing the time lag.

We can also see the three concepts (contradictions, resources, and Ideal Final Result) in the carrot seeding and latch pin examples:

1. Carrot seeding contradiction
   Tradeoff Contradiction:
   IF carrot seeds are sewn by hand
   THEN the work "tools" are simple (good)
   BUT cultivation requirements are increased (bad)
2. Resources: Biodegradable materials for fixing seed spacing, such as paper and straw, have been available for a long time.
3. Ideal Final Result: Tape for positioning seeds.

1. Latch pin contradiction
   Tradeoff Contradiction:
   IF the fit between a pin and slot is tight on a latching mechanism
   THEN the pin will not loosen easily (good)
   BUT it will be difficult to remove the pin when needed (bad)
2. Resources: Geometry has always been an available resource.
3. Ideal Final Result: The change of geometry, that is, changing the form of the pin from cylindrical to conical, is used.

Figure 5.3 presents one of the cases, carrot seeding, in visual form. Figure 5.4 is a template for presenting your own examples.

5.3 Using the Concept of Resources

Resource analysis is a handy tool with many benefits:

■ Getting new ideas directly
■ Solving contradictions
■ Predicting the system evolution

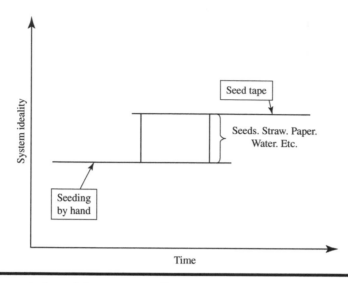

Figure 5.3   The evolution of the carrot seeding.

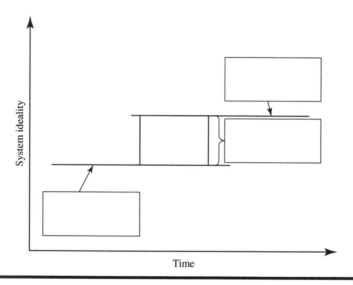

Figure 5.4   Make your own examples of how resources appear and will be used.

5.3.1 Getting Ideas

Mapping resources stimulates ideas about how to improve the system. For example, sometimes, it is enough to write down resources such as "empty space" and "geometric form" and new ideas appear. Hanging bicycles from the ceiling to be able to store more items in a garage is an example for use of empty space. Recognizing that the flat bottom of a pizza box is the principal cause of heat loss, thus making the surface corrugated can solve the problem, is an example of using geometric form.

In many business problems, the customers or users themselves are a resource. Many new types of self-service, including self-education, self-diagnosis (with home test kits for medical problems), self-treatment, self-planning (of one's house, for example), are emerging as business models that recognize the customer as a resource.

5.3.2 Solving Contradictions

If you have formulated the inherent conflicts that we considered in the last chapter, you can use resources to resolve them. Chances to get good ideas increase if you have a complete list of the resources available to solve the problem.

5.3.3 Forecasting the Evolution of Technology

When we become aware of a system's available resources, we know some features of that system's near future. Somebody, somewhere, sometime will inevitably put those resources to use.

5.4 Different Resources

Resources abound. Actually, we can make an endless list of resources if we examine the problem from greater and greater distances. How then, can we select the most useful resources? The following grouping of primary resources is one helpful technique. Groups are also illustrated in Figure 5.5.

First, divide the resources by system levels:

- Tool
- Object
- Environment
- Macrolevel system
- Microlevel system

On all system levels, a variety of resources are available:

- Substances and things
- Modified substances and things

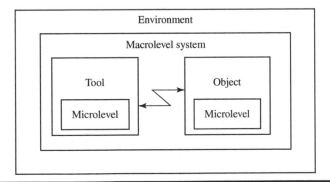

Figure 5.5   Mapping resources. List the resources found in the environment, the macrolevel system, the tool, and the object. The tool and object can contain microlevel resources and their geometry; the features and properties of the materials they are made from are resources, too.

- Voids
- Interactions and the energy to make them happen
- Form
- Features or properties
- Space
- Time

Other resources include the following:

- Information
- Harmful substances and interactions
- People's skills and abilities

You will find it helpful to examine your problem, listing everything that falls into each category. This can be called the TRIZ "treasure hunt" to find every treasure that can be used to solve the problem. This classification is a generic template, not a set of rigid rules—use it to help you start your treasure hunt. More detailed classification can be found in the literature on TRIZ, for example, in Savransky's book.[1]

Let us use the examples from the previous chapters to illustrate each of the categories of resources.

5.5 The Most Important Resource Groups

5.5.1 Resources of the Tool and the Object

The case of the latching mechanism is a typical example of the tool itself as a principal resource. In the lawnmower example, the muffler could be changed to some extent, but the most creative solution was to use a component from the macrosystem—grass. In the case of cultivating carrots, the tool was the human hand, which is difficult to change. No wonder that in this case, the object (the seed) was used as a resource, but was modified by the addition of the tape, a resource that is easily available in the environment.

The tool is often, but not always, an excellent choice as the resource to use or to modify:

- The efficiency of fighting fires has been improved by changing the tool—water. First, water was poured on fires using buckets, then pumped through hoses, then sprayed in droplets using modified hoses, and, at last, delivered as an atomized mist.
- The performance of a cutting edge has been improved by changing the materials and the geometry of the edge; by changing the geometry, the materials, and the coating of teeth (on the edge of a saw); and then by replacing the blade by powder, liquid, plasma, and laser. The function of the tool is still to cut the object, but the changes have been made to cut different materials, to increase controllability, to increase speed, and so on. The history of cutting tools illustrates the creative use of resources to respond to changing customer needs.

In business problems such as training and marketing, many tools can usually be changed. They include plans and curriculum, training, presentation materials, and the skills of the trainer. So, check the resources of the tool first.

Collect your own examples of how the tool is used as a resource (see Table 5.1).

The case of cultivating carrots is a typical example of the use of the object's resources. Many seeds are combined and can be handled as one big seed. The same concept is used when packages are combined for transportation.

Many nontechnical systems can be improved using the resources of the target group served. Fast-food restaurants, supermarkets, department stores, information websites, and innumerable other businesses that rely on self-service are perhaps the most common examples. One of the most effective ways of marketing is to deliver products and services that are so good that happy customers advertise them to potential new customers by word of mouth. Buyers themselves act as salespeople.

Collect some examples of the object as a resource (see Table 5.2).

The object cannot, unfortunately, always be used as a resource. But you should always check.

Table 5.1   List Three Examples Where the Tool Was Used as a Resource to Solve a Problem

Problem	Tool	Object	Solution

Table 5.2   List Three Examples Where the Object Was Used as a Resource to Solve a Problem

Problem	Tool	Object	Solution

5.5.2 Resources of the Environment

"Environmental resources" are things and the substances of which they are composed, energy, and the fields that always surround us. They are often ignored, because we see them every day. These resources include air, water, empty space, gravity, sunlight, and other free natural resources. They can also include the resources that are available in a specific situation. For example, most factories have available people, compressed air, electricity, information networks (sometimes computer based, sometimes paper based), and hot and cold water. Most office buildings have windows, telephone and data systems, electricity, light, and people. Hospitals have medicines, doctors, nurses, technicians, patients, beds, imaging equipment, information systems, water, air, and light. It is easy to see how something might be a tool in one situation, an object in another, and part of the environment in a third.

In northern countries in past times, ice was collected in winter, stored under a layer of sawdust, and used to cool milk and other food products in the summer. This technology became outdated when refrigerators were introduced, but it was resurrected in another form. In the Swedish city of Sundsvall, snow and ice are collected in winter and stored in a pit insulated with wood chips. The runoff is used to cool offices in warm summer months. If cold can be used as a resource, why not excess heat? In some places in central Europe, heat is collected from roads in heat accumulators and then used to keep the same roads dry in cold weather. Likewise, heat pumps use the heat available in the environment as an energy resource to operate cooling systems.

Solar energy is, of course, an environmental resource that is being widely used for many purposes. For example, solar-powered space satellites, solar water heaters, and solar-powered lawnmowers have all been available for years and span quite a range of technical complexity.

Empty space is one important invisible resource. A garbage bin partly hidden underground uses the space beneath. Empty space is sometimes used in firefighting instead of water and other substances. Firefighters attack forest fires by building fire breaks or clearings where there is nothing to burn. In a hot kitchen stove, food in the oven can sometimes catch fire when the oven door is opened. The best way to extinguish the fire is to close the door gently. When the oxygen is consumed, the fire will go out.

When solving business problems, sometimes the best thing to do is nothing. Organizations often try to improve results by emphasizing teamwork and holding many kinds of meetings. However, results often improve if the number of meetings is decreased. Many people complain that they have no time to do anything, but they may solve problems by making a decision not to do something. For example, we all know people who feel they have a moral responsibility to answer all e-mail messages immediately. An obvious solution is to throw away most messages and to answer some messages later. Time management and personal organization are popular topics for seminars and consulting. For many years, people have systematically excluded useless operations in the production of cars, computers, washing machines, and other tangible products, but there is still much "nonvalue added" work (to use a popular buzzword in the process analysis world) in the processing of information and the management of organizations.

5.5.3 The Macrolevel Resources

Any system can be combined with other systems into one greater system. Both similar and different systems can be combined. Practically everything can be improved in some respect by transition to the macrolevel. Sometimes, the effect can be dramatic. The frame and cover in a riding mower produced by the John Deere Company at one time consisted of 153 steel parts. When plastics replaced metal, only three parts were needed. The other 150 parts were integrated into the higher-level system.[2]

In the lawnmower case, one can try to resolve the conflict of big muffler–small muffler using resources on the higher system level. Grass is only one of the resources that can be used as a muffler. There are others. The casing of the lawnmower can work as a muffler. Lawnmowers are often equipped with bags for collecting grass. Obviously, the bag could be used as a muffler, too. Environmental laws are being strengthened in many regions and catalytic converters may be required for small engines. If the converter is necessary anyway, why not try to get additional benefits by also using it as a muffler?

What happens if we consider many lawns instead of one? What could that mean? The whole job could be outsourced to a service company that works on many lawns. It can use quiet electric machines that are often too expensive for small lawns. An analogy is a central vacuum cleaning system in an apartment house. Instead of a vacuum cleaner in every apartment, there is one cleaner for a whole building and only nozzles and tubes in each separate unit.

Seeds were combined into a larger system: tape with seeds. We can continue combining at higher and higher levels. What happens if we add fertilizers, soil, water, air, and other substances necessary for growing carrots to the tape? Lior Hessel, an agricultural engineer in Haifa, Israel, has developed a system that grows vegetables inside standard metal shipping containers using hydroponics. A robot-controlled system is producing 500 heads of lettuce per day, a yield 1000 times greater than a similar area can produce using conventional farming. If lettuce can be produced this way, why not carrots? Surely, there are many intermediate solutions between a tape with seeds and totally automatic farming. Revolutionary changes in gardening and farming are on the way in all parts of the world.

If you have a solution on some system level, look at higher and lower levels for other solutions. Sometimes, you will find unexpected new solutions.

In marketing and training, macrolevel resources are used even more than in technological systems. The same advertisement or the same subject matter in training is repeated many times. This is a simple combination of similar objects; modern learning theory suggests that the course should present each idea in every learning style, so that all students can benefit. Advertisers learned long ago to deliver the same message in a wide variety of channels—everything from audio to video to painting the message on a racecar or a passenger bus. Entrepreneurs are building networks all the time to provide more channels. We can view networks as resources.

5.5.4 *The Microlevel Resources*

The opposite of moving to the supersystem (macrolevel) is to go down to the microlevel. The system is segmented into smaller parts. In our example of the evolution of firefighting with water, the example of reducing water to smaller and smaller parts also illustrates the transition to the microlevel. The conflict much water–no water is solved using microlevel parts.

Microlevel parts can be empty. Pores, capillaries, holes, and the use of space are frequently called "use of voids" in TRIZ. Remember, a void is not "just nothing"—the void may have structure or texture that gives it a function to solve your problem. Textiles for outdoor clothing should be impermeable to rain and, at the same time, permeable to water from perspiration. Gore-Tex and other materials with micropores resolved the contradiction. The pores allow water molecules to pass through, but stop water drops. Foams and gels are highly technical ways of using voids (capturing "nothing" and making it do something useful). As a technical, yet old, example —whipped cream uses the chemistry of the fat molecules to capture the air and create a new product.

Segmentation may also mean that many small machines replace a big one. Imagine that, instead of a conventional lawnmower, many automatic minimowers resembling turtles are working on the grass. They may easily work on lawns with complex shapes. Distributed computing is a popular system—both within companies and on public projects such as SETI (SETI is an experiment that uses Internet-connected computers in the Search for Extraterrestrial Intelligence). The resources of many small computers on a network are used as elements of a larger computer whenever they are idle.

Let us consider the case of the latching mechanism. One can try to resolve the conflict big clearance–small clearance using different resources on the microlevel. What does this mean and how can one get there? The simplest way is to segment the pin: divide it into two parts. One of the parts may be a wedge that ensures tightness when the latch is closed. Segmentation can be continued. There may be many parts, either layers or filaments. If the segmentation is continued, more ideas appear: a pin made of powder, gel, liquid, gas, and their mixtures. How about a dynamic pin that changes its size using some microlevel effect?

Many examples of segmentation can be presented from the areas of communication, business, and education. One perpetual problem in communication is the writer's problem of sending and the reader's problem of receiving (quickly, easily, and accurately) a message that has rich content. Charles Dickens' nineteenth-century novel, *The Pickwick Papers*, was first published over 20 months in subsequent editions of a journal. It was the first serialized novel and a literary forerunner of radio and TV series in the twentieth century. All Dickens' novels were first published in parts and had enormous success. Readers found parts easier to read. Further, they got an extra benefit: a new dimension of suspense.

5.5.5 Time Resources

Problems and contradictions can often be solved by using different properties of the system at different times or by modifying the system so that it has different characteristics at different times. In carrot cultivation, the thinning operation should be and should not be. The solution realizes these requirements that, at first glance, seem incompatible. The row of plants is not thinned after the plants germinate, but it is thinned before planting.

The conical pin is tight when working and loose in the sense that it is easy to insert and remove.

Biodegradable screws, for holding broken bones in alignment, vanish when they are not needed anymore. The operation to remove the screw also vanishes from the surgical process.

Modular systems are often used to make it easy to modify the properties of a system over time. A modular bookcase can easily be made smaller or larger when the owner's needs change.

In business problems, we can seek different time resources. Sometimes, we have too little time and, sometimes, we have too much. Nearly always, some time resources can be found somewhere.

5.5.6 Space Resources

The conical pin separates properties not only in time, but in space. It is loose or easily movable in the direction of its axis, and is tight in the perpendicular direction.

In the lawnmower case, one can try to resolve the big muffler–small muffler conflict using some space available. See earlier examples using the case of the lawnmower or the bag for grass as a muffler.

In the first chapter, we used examples of the garbage bin and ax. Both were improved using space as a resource—space beneath the bin and space inside the handle.

How to use space resources in business? The resources here are different places that can be used for work and study: home, train, bus, and airplane. In conventional business, the customer is in one place and the supplier is in another. In many emerging business models, the customer is brought inside the business, either virtually (e-business systems, shared information) or physically (customers participating in product design). Other cases of spatially integrated functionality are suppliers building specialized minifactories inside their customers' facilities and specialty boutiques inside department stores.

5.5.7 Energy Resources

There are also resources of energy that are available for solution generation in any given problem. "MATChEM" is a mnemonic device to help trigger the consideration of energy fields for use in problem solving. The mnemonic stands for mechanical, acoustical, thermal, chemical, and electromagnetic fields and interactions. The general solution path is to identify which field/interaction type is enabling the function associated with the problem and consider if one of the other field/interactions could solve the problem. For example, in the latching mechanism problem, the initial problem of the loose and tight pin is based around a mechanical field/integration. Could one of the other fields solve the problem? For example, could a thermal field/interaction solve the problem? What if the latch pin was heated (to expand it while locked) and cooled (to shrink it for removal) in operation? While this solution sounds interesting on the surface, it appears the added complexity in implementing the idea might drive us further from ideality, not closer. What about the field/interaction of electromagnetism? What if the pin was a magnet? Might that improve the cylindrical pin or augment the conical pin solution? More details regarding MATChEM, and its application, can be found in Sections 7.7.3 and 10.9.

5.5.8 Other Resources

There are other important resources. All products have some aesthetic appearance and they provide information. In the ideal case, the system works well, looks good, and tells us how to use it. For example, a well-designed door tells you how to open it without labels such as "push" or "pull."

Harmful substances and interactions can be considered resources, too, sometimes called blessings in disguise. For example, waste (materials and energy) that can be used in another place or recycled are also resources. Many medicines have been developed from poisons, by finding ways to protect the healthy tissue while destroying the unhealthy parts.

Materials and technologies known in other industries can often be used to solve problems. They can also be considered resources. Microwave ovens came from radar technology. Cutting with water, Gore-Tex, remote cardiac measurement, and many other technologies and materials were the results of space research. Production technologies are slowly spreading into the construction industry. Internet resources (e.g., online class notes, course registration systems, and other applications) are spreading from business into education.

Anything that anybody else has done to accomplish a particular function is a resource for solving a problem. The better the definition of the problem, the easier it will be to find those resources. By expressing the problem in nontechnical language, you may help yourself find the resources developed in other industries. For example, a farm group was trying to find a way to dry cow manure without using conventional heating methods. However, the word "dry" kept leading them to heat. When they redefined the problem as separating liquid from solids and searched a variety of databases, they found a technology using hydrophilic molecules to carry water away from a liquid mixture. This method had been used for 55 years to concentrate fruit juices without heat.

5.6 When Resources Are in Use but Should Be Rediscovered

Sometimes, we do not need a new useful action, but we need to explain why and how harmful action develops.

At the time of prohibition, smugglers developed many clever solutions, many of which have become legends, to avoid the police. One is the smuggling of liquor by boat. Inspection of the boats never found any bootleg booze. Everybody knew that it was smuggled by sea. Bribes were not an explanation—the police were honest. What was happening?

The smugglers had a contradiction: liquor should be in the boat to run their business and should not be in the boat to avoid problems with the police. The first part of the solution was to use time resources. It was enough to get rid of the liquor just when the inspectors were on board. One obvious solution was to sink the containers underwater, combining them with some heavy thing. How could they be raised back to the surface again? The real, inherent contradiction was that the containers should be heavier than water to disappear when needed and lighter than water to reappear when needed. The extra weight should itself disappear. What could be the weight that disappears easily in the water? Something that dissolves in water? Which substances are easily available and dissolve in water? Sugar? Salt? The smugglers used table salt as a weight that dissolved and the containers surfaced at a planned time (see Figure 5.6). Of course, the police eventually learned the trick and today's drug smugglers have to use different methods.

5.7 Summary

Seek idle, invisible, free, or very cheap resources in the system and its environment.

Use all the benefits from resource analysis: new ideas, new ideas by combining tools, solving contradictions, forecasting the evolution of systems, forecasting customer needs.

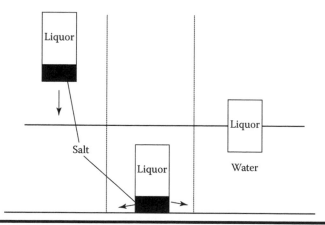

Figure 5.6 Salt and water are resources that made liquor containers first disappear and then appear again.

Map as many resources as possible, not only those that are most obvious. A deep analysis pays back. Often, unexpected opportunities can be found.

A good checklist for resource hunting is the following: tool, object, environment, macrolevel and microlevel systems, time, and space.

In this chapter, we have considered resource analysis mainly as an independent tool. In the following chapter, we will study in detail how resources are used to solve the inherent contradiction and to define the features of the Ideal Final Result. We will discuss how to select the most important (principal/primary) resource and how to use resources together.

References

1. Savransky, S. D. 2000. *Engineering of Creativity*. Boca Raton, FL: CRC Press, 83.
2. Smith, W. E. 1996. *Principles of Material Science and Engineering*. New York: McGraw-Hill, plate 2.

Chapter 6

The Impossible Often Is Possible: How to Increase the Ideality of the System

6.1 Introduction

In Chapter 2, we identified the Ideal Final Result as the solution resolving the contradiction. In this chapter, we will study the ideality of the system in detail.

We will first study the Ideal Final Result as an independent tool. We will also briefly examine different ways to describe ideal systems.

Second, we will study how to go from the definition of contradictions and resources to the Ideal Final Result. In previous chapters, we defined good solutions as those that achieve the Ideal Final Result and resolve the contradiction using idle resources. This is easy to say but hard to do. To make it easier, we need a systematic method for using resources to remove contradictions. For this, we introduce the concept of the principal resource. In Chapter 3, we described numerous different tradeoffs. In Chapter 4, we showed how to find the inherent contradiction behind the bundle of tradeoffs. In Chapter 5, we made long lists of resources from inside and outside the system. Now, we will select a single, primary, most important resource, called the "principal resource." Other resources are auxiliary. They help the principal resource remove the inherent contradiction. In this chapter, you will learn to construct the Ideal Final Result from principal and auxiliary resources. In Figure 6.1, we review the general model for problem solving.

6.2 The Law of Increasing Ideality

The concept of the Ideal Final Result is based on the law that Altshuller first formulated as follows: "The development of all systems proceeds in the direction of increasing the degree of idealness."[1]

In other words, systems get more, not less, valuable. The ideal system is the one that delivers the functionality, or more, of the original system, but has no weight and no volume, requires no

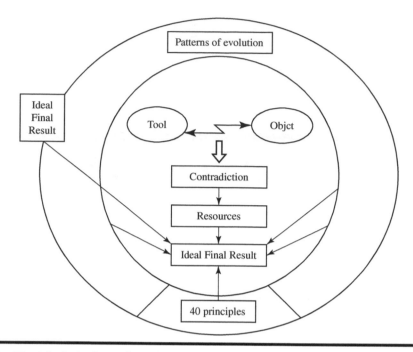

Figure 6.1 The Ideal Final Result in the model for problem solving. The arrows show that the Ideal Final Result can be developed by the application of many different tools.

labor and no maintenance, consumes no energy, and so on. Where does this definition come from? Start with the ideality equation:

$$\text{Ideality} = \frac{\Sigma\,\text{Benefits}}{\left(\Sigma\,\text{Costs} + \Sigma\,\text{Harm}\right)}$$

The Greek symbol Σ means "the sum of," so this equation reads, "Ideality is the sum of all benefits divided by the sum of all costs and all harm." Examples of harm can be the damage done to other systems, the waste a system creates, or the injury it inflicts on people (a subset of the first in the list). If all harm associated with a system is monetized, then the equation's denominator can simply be represented by Σ Costs. If you dislike formulas and equations, do not worry. The formula is qualitative. We will not do any actual calculations. The point of the formula is that it clearly illustrates two sides of ideality.

The formula generalizes numerous expressions presented to describe the level of technologies, inventions, and solutions. It was adapted from the value equation in *Techniques of Value Analysis and Engineering* in the early 1950s[2]: Value is the capability of the function divided by costs ($V = F/C$). Benefits can contain functional capabilities, but are not limited to them. Many important features, such as weight and size, are not actions or functions.

In his book *Great Inventions through History*, Gerald Messadié compares inventions with the fishing technique of the seagull:

> The seagull, which carries a clam in its beak, places it on a wall, then goes to pick up the biggest stone it can manage and drops it from a height in order to break the shell, has invented the technique which is inspired by neither the spirit of

commerce nor the desire for power. This technique enables it to obtain its food reasonably quickly and in return for a little ingenuity. ... Thus, it saves time and consequently energy: this is the goal of absolutely all the inventions which have been made since humanity began. ... All inventions are included in this absolute rule.[3]

The rule formulated by Messadié stresses very important features, but is nevertheless too narrow: The increase of ideality could be a faster, more energy-efficient system, but lots of other changes are possible, too, such as decreasing weight and size, improving outer appearance, increasing comfort, or decreasing harmful by-products, for example. Nevertheless, Messadié's rule is not a bad illustration of the law of increasing ideality. Examples can be found easily. Observation of our world shows that time or energy or both are saved in many improving systems, from those as simple as our carrot-planting example to the development of computers and transportation systems. Cars and airplanes are means of traveling more quickly and with the least possible effort. Electric lighting is extremely easy to use compared with torches and earlier lamps of any kind. The same can be said of food processing, communication technology, and all other technologies that are widely used—their commercial success is proof that they have made life easier for people.

The maximum value of the equation is reached when the denominator is zero—that is how we concluded that the Ideal Final Result is a system that achieves all the benefits with no cost and no harm. At first glance, having all (or even most) cost and harm disappear from a system is difficult to comprehend. For cost and harm to disappear, then the system's components would have to also disappear. How can a system still provide a function if there are no (or substantially fewer) components associated with it? Consider the example of the ideality transition of the kerosene lamp to the modern electric light. The light goes from an entire operating system (kerosene lamp) to a partial operating system (light bulb). Where did the rest of the system go? This example represents one of the patterns of system evolution (Chapter 10), the trend of transition to the macrolevel (supersystem). Referring to Figure 6.2 it can be seen that the fuel, fuel reservoir, fuel transmission, and fuel control all transition to the macrosystem (supersystem) as the power plant, transmission lines, and

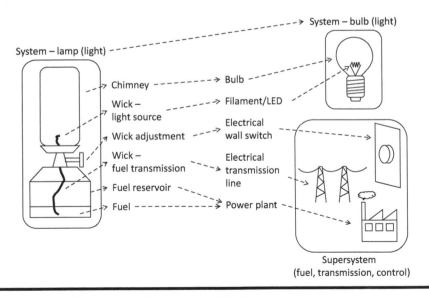

Figure 6.2 Kerosene lamp's transition towards ideality.

electrical wall switch. The most important component of the kerosene lamp (the light-emitting device) is the only thing retained in the new system (the light bulb). Because complexity causes an increase in cost and harm, increasing simplicity will increase ideality if the benefits stay the same.

Ideality is measured by comparing systems. We can easily say which of two alternative systems is closer to ideal (when holding constant a specific set of circumstances applied to both systems) by comparing the ideality equation. If benefits increase with no change in cost or harm, then ideality increases. If harm is reduced with no change in benefits or cost, then ideality increases. Some examples:

- In the 1970s, the pocket electronic calculator quickly replaced the slide rule. The calculator was more precise, easier to carry, and easier to remember how to use. When the price of the calculator fell, the slide rule quickly became a museum piece.
- Several rotating piston engines for cars have been developed in an attempt to find an alternative to the conventional internal combustion engine. Ideas for replacing the linear motion of pistons with rotary motion were developed early in the history of the steam engine—James Watt got a patent for a rotating steam engine in 1769. Rotating machines have their appeal. Steam and gas turbines are good examples. Nevertheless, a rotating engine has not superseded conventional ones in cars. The benefits are too small compared with the drawbacks, such as sealing problems and the high consumption of fuel. A clear increase in ideality and a clear supremacy over the competing system are absent.
- Scissors formed to fit the human hand (Fiskars and some other brands) have become popular and partially replaced conventional models. A small change of geometry produced a noticeable improvement in comfort.
- Telephones and many other products today have much simpler, less complex forms than they did early in their development.
- Corporate logos and fonts in typography have become simpler, too. Although fashion may be at work here as well as the evolution of systems, the same message is delivered using simpler and leaner forms.
- Control of automobile braking and steering systems "by wire" instead of by direct mechanical linkage is 20 years behind the development of by-wire systems for aircraft. However, it is advancing quickly and in the direction of increasing ideality. The new system can stop a car in an emergency in 10% less distance than conventional systems, using 15 parts instead of 45.[4]
- The first long-distance communication system was a signal fire. Next was a series of signal fires, each lit as the lookouts saw the previous fire lit. The next evolution of the system was horns and drums. There are other potential candidates for the listing, but one path could be speaking tubes on ships, the telegraph, the telephone, the cordless phone, the mobile phone, and then the Internet. As can be seen, many of the transitions involved what is referred to within innovation circles as "changes in principle of action."

Sometimes, scholars who view technology as one part of the social organism see important features better than engineers. Historian Arnold J. Toynbee gives some excellent illustrations from the evolution of technology and science, where he sees "a law of progressive simplification." First, transportation:

> When the horse was replaced by the locomotive, the simple carriage-road had to be turned into an elaborate 'permanent way,' with ... a pair of metal rails ... in the next stage of technical advance, when the ponderous and bulky steam engine ... is replaced

by the light and handy internal-combustion engine, ... the improvement in technique is accompanied by a notable simplification of apparatus. ... The technical advantage of mechanical traction is not only preserved but enhanced (inasmuch as the internal-combustion engine is an improvement on the steam engine from the mechanical standpoint); and at the same time the disadvantage of the elaborate material apparatus is partly transcended. For the motor-car liberates itself from the rails to which the locomotive is bound and takes to the road again, with all the speed and power of a railway train and almost all the freedom of action of a pedestrian or a horse.[5]

Keeping in mind that not all transitions toward ideality involve simplification, it can be seen that the steam engine, which is vastly more complicated than the horse, was still an improvement of the overall value of the transportation system. Then, the automobile made the next step toward ideality and was somewhat less complicated than the steam engine, thus increasing the overall value of the system once again. We are now in an age of the electric vehicles that have thousands of fewer parts than the internal combustion engine equivalent and can greatly improve handling and safety by way of four separately controlled direct-drive digital motors. The modern electric vehicle again pushes the transportation system toward increasing ideality. Once again, we can see a definite change in principle of action as the system moves from horse to steam engine, to fossil fuel automobiles, and, then finally, to the electric automobile.

Toynbee also has illustrative examples of the evolution of telecommunication, writing, fashion, and astronomy. The telegraph and telephone were invented first and transformed business and society with a speed of communication not imaginable earlier. More recently, wireless communication has dramatically improved the transmission and accessibility of information.

Fashion has evolved from the extravagant costumes of Queen Elizabeth I or King Louis XIV toward plainer materials and simpler cut. The triumphal march of denim and casual clothes, even in traditional businesses, has continued the trend in this century.

Astronomy and physics have long recognized the usefulness of increasingly simple models and theories. The Ptolemaic (earth-centered) system had to postulate complex epicycles to explain observed movements of known heavenly bodies. The Copernican (sun-centered) system presents in far simpler terms a wider range of movement of innumerable bodies. Modern theories of elementary particles use six (or so) quarks to describe all other particles—in the 1960s, more than 200 particles were used to describe matter.

Let us consider some typical ways to improve systems. If the useful features are clearly much improved and the greater numbers of parts cause very little harm or cost, increasing the number of parts can improve ideality. A modern bicycle has more parts than the first hobbyhorses in the late eighteenth century. The features are improved so much, however, that the ideality of the bicycle is increased. Overall, over history, the number of parts will decrease. The new electronic gearshift mechanisms are much simpler than the mechanical shifting systems that have dominated bicycle technology for the past 50 years.

Another way to increase the ideality of a system is to decrease the size, energy consumption, weight, the number of parts and operations, and other cost-generating factors. The frame of the bicycle is usually made of a few parts. A single-part frame is used for sport bikes. The ideality is improved by decreasing the number of parts. In this case, ideality was improved based on a specified goal. If the goal is different, the ideality may or may not be improved.

Instead of the term "ideal system," we can use terms such as "ideal machine," "ideal process," and "ideal substance" (or "ideal material"). The ideal machine does not exist, but its job gets done. The ideal process is one that consumes no energy and no time, but produces the desired product or service. The ideal substance is one (nonexistent) having all needed features. Here are examples of each variation:

- I want to produce parts, but I do not want to build a factory with lots of specialized machine tools. The current solution, which is on the way to an ideal solution, is the stereo lithography system that can produce parts from a three-dimensional CAD file and a selection of metal, plastic, and ceramic powders.
- I want to go to exotic places, but I do not want to spend time traveling. Old solution: read a book. New solution: visit websites that have real-time cameras in exotic places. The Star Trek solution—instantaneous travel—is impossible now, but it may be possible in the future.
- I want clean clothes, but I do not want to do the work of washing them. Two pathways exist up this "mountain." Along one path are services that take your clothes, wash them, and return them without any work on your part (but the cost is not zero). Along the other path are machines that wash and dry clothes with increasing sophistication (sorting, selecting the right temperatures and wash methods, etc.), but the machines and the chemicals also have nonzero cost. The next step on this path, currently being tested by several manufacturers, is the incorporation of ozone-generating materials in the machine so that soap or detergent will not be needed, reducing the operating cost and reducing the harm that the wastewater does to the environment.

In the ideal system, the harmful feature disappears and the useful one is retained. The solution gets closer and closer to the ideal if the harm can be turned into a benefit or a blessing in disguise, as scrap merchants have traditionally turned waste to profit. Later design for remanufacture, design for recycle, and, generally, design for the environment illustrate increasing ideality. Examples are the following:

- Biological waste can often be converted to valuable and environmentally friendly biogas.
- Wastepaper becomes raw material for recycled paper and packaging products. For this reason, experts joke that the biggest producer of paper in the world is New York City.
- Customer complaints are certainly undesirable, but if a company learns from the complaints and creates a better product, then the harm turns into good.

Add your own examples to this list (Table 6.1).

Table 6.1   List Examples Illustrating Increasing Ideality in Systems with Which You Are Familiar

Initial system	
Improved system	
What changed?	
Benefits improved	
Cost reduced	
Harm removed	

6.3 Constructing Solutions from Resources

Sometimes, the formulation of contradictions and the mapping of resources nearly directly tell us how to solve the problem. If we know that the garbage bin should be small *and* large and that geometric space is one of the resources, we can rather easily discover the idea of using the space beneath the bin. The visible bin remains small and the bin as a whole gets large.

The information or resources may, however, not be enough to find an idea for the solution. In Chapter 5, we considered the smuggling problem. We know that the canisters should disappear at certain times and appear again at another time. Analyzing contradictory requirements for the tool, we concluded that the canister should be heavier *and* lighter than water. Then, the canister can be dropped temporarily into the water and sink under the surface, but later float at the surface for easy recovery. Resources are a canister, liquor, water, buoyancy of water, gravitation, and resources of macrolevel systems. This information may, however, not be enough to find an idea for the solution.

Additional steps are needed to move from the resource analysis to the Ideal Final Result. First, we select the principal resource. Remember that the principal resource is the primary, most important resource, or the resource that exhibits the inherent contradiction. This primary, or principal, resource is chosen because it is the resource that is closest to the problem and is, therefore, most likely to be able to make contributions toward solving the problem at hand. It is important to keep in mind that a resource can be a physical thing or a parameter of the physical thing, or both. For example, boiling water is a physical thing, but it may be the parameter of 212°F, or 100°C that is the utilized resource. If you have defined the inherent contradiction, described in Chapter 4, it is rather easy to find the principal resource. Recall some examples:

- In the smuggling problem, the canister should be heavy *and* light. The principal resource is the canister.
- In the lawnmower problem, the inherent contradiction is big muffler–no muffler. The principal resource is the muffler.
- In the carrot-cultivating problem, the inherent contradiction is many seeds–one seed. The principal resource is a seed.
- The problem of the latching mechanism: the contradiction is no clearance–big clearance. The principal resource is the pin.
- In the training problem, lots of time is needed for training and no time at all is available. Time is the principal resource.
- In the firefighting example, we need substantial water and no water. Water is the principal resource.

Auxiliary resources can change the principal resource so that the contradiction disappears. The smugglers found an excellent auxiliary resource: salt, making the canister first heavy and then light again. In the lawnmower problem, grass helps make the muffler smaller. In the carrot-cultivation problem, something connecting seeds helps make one seed from many seeds. In the example of the latching mechanism, the geometry of the pin itself makes the clearance change from big to zero. In the training example, working time is the resource that can be used to minimize training time. In the example of firefighting, high pressure makes substantial "water" from almost none (mist).

These examples are simplified, of course. Analysis of the whole situation is needed to find proper auxiliary resources. Smugglers, obviously, used some time to figure out that there was salt available. It is useful to list more than one auxiliary resource. Sometimes, changes of the principal resource are needed to get a good solution. List all the attributes of the principal resource to find candidates for change. In the case of the canister, for example, buoyancy was the attribute that was changed.

To review the formulation of the Ideal Final Result using resources requires three steps:

1. Select the most important or primary resource having an inherent contradiction. See Chapter 4 on intensifying contradictions for help.
2. List auxiliary resources or resources that can change the primary resource. See Chapter 5 on resources.
3. Change the principal resource by using auxiliary resources so that the contradiction vanishes.

These steps can be conveniently organized in a table (Table 6.2) and illustrated using our examples. Many auxiliary resources will make other solutions possible.

In the noise problem, an exhaust tube, exhaust gases, and grass make the big muffler small or, even better, make a big muffler into an absent muffler (Table 6.3).

If we use different resources, we will get different solutions. For example, if the casing of the lawnmower is used as an auxiliary resource to redirect the sound and to absorb it, the case, instead of the grass, becomes the muffler.

In the example from carrot cultivation, seeds, soil, water, and other resources change the seeds so that the number of objects is, in some sense, large and small at the same time (Table 6.4).

Table 6.2 Constructing the Ideal Final Result in the Smuggling Problem

Primary resource with the inherent contradiction: canister
High density–low density
Auxiliary resources: water, air, salt, sand, sugar, etc., time, gravity, buoyancy
Features of the Ideal Final Result: salt changes the canister—heavy (high density) and then light (low density)

Table 6.3 Constructing the Ideal Final Result in the Lawnmower Problem

Primary resource with the inherent contradiction: muffler
Big muffler–no muffler
Auxiliary resources: grass, exhaust tube, exhaust gas, air
Features of the Ideal Final Result: grass makes muffler present and absent at the same time

Table 6.4   Constructing the Ideal Final Result in the Carrot Cultivation Problem

Primary resource with the inherent contradiction: seed
Many seeds–no seeds
Auxiliary resources: soil, water, wastepaper (from food packages), straw, mulch
Features of the Ideal Final Result: tape made from wastepaper and other cheap materials combines many seeds into one seed

Table 6.5   Constructing the Ideal Final Result in the Example of the Latching Mechanism

Primary resource with the inherent contradiction: pin
Big clearance–zero clearance
Auxiliary resources: pin: geometry, surface, material, time
Features of the Ideal Final Result: geometry makes clearance wide when the latch is open and zero when it is closed

Table 6.6   Constructing the Ideal Final Result in the Training Example

Primary resource with the inherent contradiction: time
Lots of time–no time at all
Auxiliary resources: working time, existing knowledge and skills, the culture of the company, curriculum, textbooks, computer networks, students, experienced people, teachers, etc.
Features of the Ideal Final Result: work has a training effect. Training takes place over extended time periods, without special training time, so that training gets better and work results get better, too

In the example of the latching mechanism, the geometry of the pin is a resource that makes the clearance both large and small (Table 6.5).

In the training example (Table 6.6), working time is used to get plenty of time when there is actually no time available. Work can be a form of on-site training if it is well designed, with lots of feedback so that the worker can learn from each experience. The longer employees work, the better trained they are. The solution is unconsciously used all the time.

Consider one final example of firefighting (Table 6.7).

Examples help you to study your own system (see Table 6.8). Continue the exercise in the previous chapter and study how the resources you have mapped can be combined. You can also use the concept of ideality directly—decide what are the primary and auxiliary resources and how they can be used together.

Table 6.7 Constructing the Ideal Final Result in the Firefighting Example

Primary resource with the inherent contradiction: water
Much water–no water
Auxiliary resources: water, water pressure, tubes, nozzles, air
Features of the Ideal Final Result: high pressure makes the amount of water nearly zero and very large—the volume of mist is very high

Table 6.8 Study Your Own System

Primary resource with the inherent contradiction:
Auxiliary resources:
Features of the Ideal Final Result:

6.4 Summary

Study the difference between good and weak solutions. Increasing ideality is one important feature of good solutions. Numerous examples show that systems really can get simpler, even though they solve complex problems. Big benefits can be created with low cost and little harm. You can use the concept of increasing ideality directly—study good solutions in other industries and you will get ideas about improving your system.

If you have defined the inherent contradiction (Chapter 4) and mapped resources (Chapter 5), you can very effectively build the Ideal Final Result from resources. Select the system with the inherent contradiction as the principal resource. Find auxiliary resources that can be used to change the principal resource so that the contradiction disappears.

The features of the Ideal Final Result form the basis of the method for evaluating solutions. The evaluation and improvement of solutions are considered in Chapter 8. Different methods can increase the ideality of the system, for example the system can be trimmed (see Section 7.6.2) and the patterns of system evolution can be applied, as studied in Chapter 10.

References

1. Altshuller, G. S. 1984. *Creativity as an Exact Science*. New York: Gordon and Breach, 228.
2. Miles, L. D. 1961. *Techniques of Value Analysis and Engineering*. New York: McGraw-Hill.
3. Messadié, G. 1991. *Great Inventions through History*. St. Ives, UK: Chambers, 5.
4. Turrettini, J. 2001. Wired wheels. *Forbes*, 168:3, 85–86.
5. Toynbee, A. J. 1963. *A Study of History, Vol. III*. London: Oxford University Press, 174.

Chapter 7

Understanding How Systems Work: Utilizing Functional Analysis to Expand Knowledge About Your Problem

7.1 Introduction

In previous chapters, we have shown how to establish desirable solution directions (Chapter 2), develop tradeoff (Chapter 3) and inherent (Chapter 4) contradictions, look for resources that can support solution concepts (Chapter 5), and improve the value of a system (Chapter 6). In this chapter, we will explore a powerful tool for gaining deeper insight into the system in need of improvement by way of a method called "functional analysis."

This chapter utilizes several different functional models as learning aids. Some are somewhat straightforward, while others are rather complex. It is not necessary to fully comprehend every functional model to benefit from this chapter. Even utilizing the following methods for the creation of less sophisticated functional models will still provide insight to the problem solver.

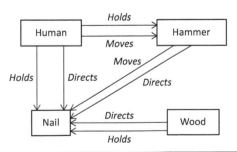

Figure 7.1   Hammer functional model.

At the most rudimentary level, functional analysis primarily utilizes a problem-modeling tool (functional modeling) that shows the elements (components) associated with a system and what the functional relationships are between those components. What is a function? A function is what gets done. It is the performance of an action such as movement, shaping, heating, or informing.

Why do we want to model systems functionally? There are several advantages to functional modeling and analysis:

1. The method provides insight as to how the system actually works in enabling the basic job (function) the system was designed to perform (e.g., car *moves* people).
2. Functional relationships show what is important about the system in that it is the functions that matter, and, therefore, our mental inertia is broken on realizing that other, modified, or fewer components can support the same functionality
3. The modeling exposes disadvantages in the system that can be addressed through contradiction analysis (Chapters 3 and 4) and other methods.
4. Every component appearing in a functional model contains resources (Chapter 5) whose characteristics (such as size, shape, position, density, insulation properties, energy) might help us to solve problems within the system and move the system closer to being an ideal system (Chapters 2 and 6). More specifically, this ties back to earlier chapters on ideality in that functional analysis supports the improvement of systems. It helps to exhibit what within the system needs to be improved and which components can be eliminated while doing so.

Functional modeling was not originally developed as one of the classical TRIZ tools, but it is so useful in providing insight into possible system improvements that it has been developed over the years into a primary problem-modeling tool, supporting the usage of other TRIZ tools. The functional analysis tool finds its primitive beginnings in the early 1900s when efficiency analysis was first being developed. Then in the late 1940s, an engineer at GE, Larry Miles, began the development of what today is known as "Value Engineering" (VE).[1] As part of his VE methodology, Miles utilized a process of using verb/noun word pairs that he called "functions."[2] As an example, a word pair could by turns shift or move information. By 1947, the basic "Value Analysis Functional Approach" was developed.[3] Today, the methodology is much more mature and is most often represented graphically where Miles' word pairs are assigned to graphical representations of actions taken on components by other components inside and outside of the system under analysis. The current state of functional analysis, as orchestrated with TRIZ, is a powerful and enlightening methodology. Functional analysis and TRIZ have each enriched the other. TRIZ is even more powerful when supported by the systems engineering and hierarchical insight provided by functional analysis, and functional analysis becomes an innovation agent when supported by the science of TRIZ. For more advanced topics on functional analysis, see the paper by Litvin *et al.*[4]

Figure 7.1 provides an example of a simple functional model of a hammer. A more precise description is that Figure 7.1 shows a hammer in its operational environment. A functional model of a hammer simply lying on a workbench would be rather uninformative (workbench holds hammer). But at least we would understand that when storing a hammer, the important function (or action) is "holds," and we can then think of other ways to hold a hammer for storage. In Figure 7.1 it is shown that even in a simple four component functional model, there are many more functions (eight in the hammer example) than there are components. As shown, most components perform more than one function (or action) within, or around, the system.

7.2 Functional Language

Before delving into the details of creating and utilizing functional models, it is necessary to understand the particulars of functional language. Functional language is semantic in nature in that it utilizes subject–action–object relationships to describe the interaction between two or more physical elements (components). Ink marks paper is an example of a subject (ink) action (marks) object (paper) relationship. Others are pipe directs water, mirror informs person, computer memory holds data, belt turns pulley, delivery person moves parcel. An important aspect of the selection of function (action) words is that they should represent the fundamental action being performed, utilizing simple language, not technical terms (i.e., jargon). For example, as opposed to saying "plasma cutter vaporizes metal," it is better to say "plasma cutter removes metal." While the plasma cutter may very well vaporize the metal in removing it, the real goal is to remove the metal, not vaporize it. If the term "vaporize" is used, then when looking for alternative methods (if that is indeed necessary toward the problem solving) only vaporization systems will be considered as opposed to a more expansive set of removal systems. Further, jargon carries with it specific meanings that not only can falsely specify a function's requirements but also can confuse others who work in different fields or disciplines. In a business environment, analogous examples include moves versus transports, arranges versus collates, or informs versus educates, with the former of each word pair being the more direct, clear, meaningful, and, therefore, better functional word choice.

Take a few minutes and capture a few functional statements of your own in Table 7.1. Keep in mind that the subject (tool) provides the action (function) and that the object (recipient) is the item (component) being affected (changed).

7.2.1 How Are Functions Defined?

More technically, a function is an action of one material object (item) on another, which changes or maintains a parameter of the affected object. A parameter is a measurable factor that supports the definition of a system. The parameter can either be represented as a numerical value (e.g., 93°C or 199.4°F) or not (e.g., temperature). Therefore, motor moves rocket describes a function because the three-dimensional location (parameter) of the rocket is changed by the motor. The motor creates a change to the rocket.

Table 7.1   Functional Statements Exercise

Subject (tool)	Action (function)	Object (recipient)

The key to functional language is to describe what is actually happening in order to perform the function. Sometimes, this is not quite as obvious as might be expected. As a case in point, consider an open door. What might be the functional description of the action between an open door and a person? Person uses doorway? Door allows person to walk through it? Door allows passage? As we can see, it is difficult to describe the function of an open door. The reason is that when a door is open, it has no function to describe. Only when the door is closed does it perform a function: door stops person. When the door is open, it might as well not exist. An archway is just as useful as an open door in that situation, but then the archway, as with the open door, has the function of directing people (i.e., shows where the wall can be penetrated—through the opening). The same is true of electrical switches and water valves. When a switch is open, the switch provides no functionality. In other words, there might as well just be a wire in its place to conduct the electricity. So, what is the function of a switch? Switch stops electricity. What is the function of a valve? Similarly, there is no function if the valve is open. Like an electrical switch, when the valve is open it might as well just be a pipe. Just like the switch, the valve only functions if it is closed. Granted, the valve could only be partially closed (in which case its functionality could be replicated by a smaller diameter pipe), but the function is still the same. Valve stops water.

While functional language has roots originating from written and oral language practices, it differs a bit in the rules by which it is properly applied. Recalling from Section 3.4.1 where the action between a tool and its object is described, it was stated that "Action means that the tool does something that causes the object to change or to be maintained." This is the subtlest aspect of the three requirements for a functional relationship to exist:

1. There are two material objects
   (Physical "things" such as nails, documents, neutrons, sound waves)
2. That interact with each other
   (Physically come into contact with one another—it is important to understand that sometimes the energy field (see MATChEM from Sections 5.5.7, 7.7.3, and 10.9) can be ignored. For example, siren informs person; clearly the siren is not physically touching the person. A more precise description, though not always necessary, would be siren creates sound waves, followed by sound waves inform person.)
3. And the acting object changes or maintains a parameter of the recipient object
   a. Parameter: A comparable value of an attribute of a material object position, shape, temperature, knowledge, etc.
   b. Changes or maintains: Change is fairly easy to understand—truck moves cargo, knife removes material, teacher informs student. However, maintains is a bit more perplexing until its intent is understood. Many functions maintain parameters of their recipient objects with the most common maintenance function being that of holds: nail holds painting, table holds dishes, bearing holds shaft. Another maintenance function example within a business analysis could be coach reassures athlete.

Therefore, if two material objects come into contact with each other and the acting component changes or maintains a parameter for the recipient object, then a function has occurred. Once again, this appears simple enough, but there are a few scenarios where it is not as straightforward as it sounds. An example, in regard to a technical system, is the measurement of a signal. An oscilloscope or a multimeter is used to measure a signal. Therefore, the functional relationship might be expressed as meter measures signal. Does this functional relationship meet the three requirements of a function? Requirement one states that there are two material objects. Yes, a test meter and an electronic signal (moving electrons) are both material objects. Requirement two states that the objects must come into contact with each other. Once again, the answer is yes. For the meter to register, the signal they must be in contact. Requirement three states that the acting physical object (meter) must change or maintain a parameter of the recipient object (signal). Does a test meter change a signal when measuring it? Electrical engineers are taught that you cannot test an electronic system without changing it. For example, testing an electrical current draws off a small bit of that current in order to trigger the meter, and this changes the current being tested. However, that is not the purpose of our system, just an undesired side effect of the test. Therefore, in measuring a signal, a meter does not change the signal. Then, what is the actual function and how do we express it? What is really happening in testing a signal? The purpose of testing a signal is to gain information about the signal and display or record that information. Therefore, the desired function is not meter measures signal but rather signal informs meter. There is a very similar condition in business systems in regard to information exchange. For example, does a hiring manager interview a job applicant? In testing that relationship with the three requirements of a function, it can be seen that requirement one, of having two physical objects, is met. Both the hiring manager and the job applicant are physical objects (persons). Requirement two is also met in that the hiring manager and the job applicant are in physical contact with each other, at least in the case of a face-to-face interview. More correctly, there are sound waves that actually do the contacting (even if it is a phone interview), but, as we've discussed, sometimes the energy carrying the function can be ignored. Requirement three is not met as the interviewer does not change or maintain something about the job applicant. Rather, the job applicant informs the interviewer. What is being changed is the interviewer's knowledge and understanding of the job applicant. Along the same lines, a person does not read a document (what is changed or maintained about the document when a person reads it?—nothing). Rather, the book informs the person. The person's knowledge is changed by the document.

7.3 Function Types and Quality

The previous *motor moves rocket* example demonstrates a useful type function assuming that the parameter change of the rocket (its location) was desired. However, there are also harmful-type functions. Harmful functions are those that result in parameter changes that are undesired. For example, nail damages tire expresses a harmful function because it is not desired for the nail to damage the tire. Therefore, useful functions change (or maintain) parameters in a desired direction, while harmful functions change parameters in an undesired direction. Referring to the lawnmower muffler example (see Section 3.4.4), the muffler performs both a useful and a harmful function. Muffler absorbs noise is a useful function, while muffler slows exhaust gas is a harmful function as the slowing of the exhaust gas creates engine back pressure that, in turn, reduces the efficiency of the engine. Digging in a bit deeper, it can be shown that there are three qualities of useful function performance: sufficient, insufficient, and excessive. A sufficient useful function changes a parameter exactly as much as desired. For example, in Figure 7.2, in relation to the function—person throws ball—throwing a ball exactly to the other person is a useful and sufficient function. The desire was to throw the ball to the other person and the result was exactly that. On the contrary, if the ball is thrown short of the other person, then a useful function has still been performed, as the ball was moved in the desired direction, but it was insufficient. And further, if the ball is thrown over the head of the other person, then a useful function has once again been performed but it was excessive. Finally, if the ball is thrown in the wrong direction altogether then a function is still performed,

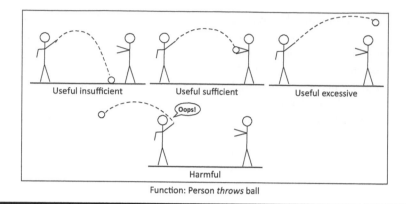

Figure 7.2   Useful and harmful functions of throwing a ball.

but that function is making changes to the ball that are not desired in support of the desired final result. Therefore, throwing a ball in the wrong direction would be considered a harmful function. Functional models represent the functions' categories (useful, insufficient, excessive, and harmful) by way of different styles of function arrows, as shown in Figure 7.3.

7.4 Component Analysis

To build a functional model, it is necessary to decide upon the components that will go into the model and what those components' relationships are. It is necessary to choose components for the analysis at the correct, and same, hierarchical level. If a mobile phone manufacturer wants to innovate the interface between the person and the mobile phone, then they could start by functionally modeling that interface. A functional model of a person using a mobile phone would most likely include components such as a screen, case, speaker, microphone, ear, mouth, and finger. Why not throw silicon-based transistor into that list? The silicon-based transistor is not at the user-interface hierarchical level nor is the cell tower handling the phone's signal. It is best to ensure that all components considered are at the same system "level" in relation with each other.

Another important point is that duplicates of the same component only need to be listed once. For instance, if 10 identical bolts secure a valve cover to a manifold, the model only needs to include the bolts together, not each one separately. Only if there are two or more types of bolts (e.g., securing bolts and grounding wire bolts) would more than one bolt type be listed. Similarly, for a business analysis, multiple clerks at the motor vehicles division only need to be listed once (assuming they all do the same job). The same is true for file cabinets. Regardless of how many file cabinets are in a work space, if they all hold files then the file cabinets only need a single listing as a system component. If, on the other hand, some file cabinets hold documents and others hold reference books, then it might be better to list two kinds of file cabinets: document files cabinets and reference file cabinets.

A functional model can be constructed from as few as two components (poison kills fly) or hundreds of components. This was the exact scenario when the chapter author used functional modeling to help improve satellite reliability for the US Air Force. It was necessary to model dozens of components to represent the satellite, and many others were needed to understand the environmental threats to the satellite. The good news is that generally a functional model should have somewhere between 7 and 15 components. More or fewer is also acceptable, but the 7—15 rule helps to ensure that the system is being analyzed at the correct hierarchical level.

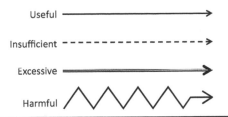

Figure 7.3   Styles of arrows used to represent function categories.

Component analysis is the process of listing the system components and then indicating whether they come into contact with each other or not. Understanding the component interconnections indicates which components will have functional relationships between them. One of the easiest ways to execute this analysis is by way of an interaction matrix. Simply list the components of the system under analysis on the X and Y axes of a matrix. Second, place a "Yes" ("Y" is fine) in the intersecting boxes for each pair of components that physically touch each other. The example of egg mixing can be seen in Table 7.2. The components involved in the mixing of an egg have been listed, and it is then indicated whether they touch each other, or not. The "touching" is indicated by a mark in the intersecting rows and columns. By the way, it should also be noticed that the table should be symmetrical about the 45-degree axis. In examining the matrix, it is shown that all of the components have an interaction with all of the other components except for the hand/egg pairing. If the egg mixing is done properly, the egg and the hand should not come into contact (see Figure 7.4).

7.5 Functional Modeling

To perform a functional analysis, a functional model is required. As discussed previously, a functional model is a model of a system that shows the elements (components) associated with that system and what the functional relationships are between those components. It is important to keep in mind that a functional model does not contain any information regarding the timing of the actions (functions) reflected in the model. With the example of a blacksmith's work, the blacksmith would first heat metal and then form the metal. Both actions would be reflected in the functional model of the system, but there is no indication as to which comes first or rather even

Table 7.2   Component Analysis of Egg Mixing

Egg-mixing interaction matrix	egg	fork	hand	bowl
egg		Yes		Yes
fork	Yes		Yes	Yes
hand		Yes		Yes
bowl	Yes	Yes	Yes	

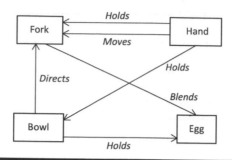

Figure 7.4 Egg-mixing functional model.

if they occur at the same time. In functional modeling, timing is irrelevant. What is important about a functional model is what components are involved and what the functions are that they perform on each other.

7.5.1 Building a Functional Model (Blacksmith Example)

In most cases, a functional model is built as a tool to support the improvement of a system. If improvement is needed, then there is most likely some problem that has been identified. The model should, therefore, reflect the "problem" within the system. Further, the person building the model needs to be intimately familiar with the system under analysis, or at least have technical support from others who do. The following is an example of building a functional model of a blacksmith's simple metal-forming operation.

Any component necessary to the operation of the system should be reflected in the associated functional model. In the case of the blacksmith's metal-forming operation, the components might include the blacksmith, glove (protects hand), workpiece (metal rod), fire (for heating workpiece), tongs (for holding workpiece), anvil (workpiece support), and hammer (workpiece shaping). In other words, think through what tools, machines, humans, materials, and equipment are necessary to the completion of the task(s) under analysis.

As discussed, there is likely a problem to be addressed whenever a functional model is being utilized. In this case, the problem will be assumed to be that the workpiece cools too rapidly during the shaping process. What would make the workpiece cool too rapidly? Heat transfer occurs when there is a temperature gradient between items. Therefore, it should be captured that the air around the workpiece is rapidly cooling the metal (convection). From that, it is understood that air should also be listed as a system component. While conduction between the workpiece and the anvil is also partially responsible for the workpiece being cooled too rapidly, that relationship has been ignored in this analysis.

It should be noted that when causes of problems are unknown, it is important to ascertain what is driving an issue. There are different methods that can be applied to determine why problems are occurring such as the following: ask an expert, perform an experiment, use root cause or cause-and-effect analysis, or simply make an educated guess. If the cause of a problem is not understood, it is unlikely that a useful solution will be developed.

Referring to Table 7.3, the blacksmith's work components have been listed on the X and Y axes of the table, and there is an indication as to which of these components physically touch each other. In examining the table, it can be seen that the air, which clearly touches everything, is only shown as touching the workpiece (metal rod) and the fire. The problem states that the workpiece is cooled too rapidly during working. Therefore, the model is only concerned with the air's relationship to that cooling process and its relationship to other potential heat sources (could the fire be a possible resource in the solution generation?). It would be okay to show the air touching everything else in the system; however, it was not shown as such in this example as it would clutter this introductory functional model. Now that there is a listing of the system components indicating which components physically interact with each other, it is understood which components have functions between them. In studying Figure 7.5, the component analysis, shown in tabular form (Table 7.3), has been turned into a graphical model. All the components are represented and a single, double headed arrow between components shows the interactions reflected in Table 7.3. The actual functions occurring between the components are not yet explored, but it is clear between which components the yet-to-be captured functions will operate.

It is now possible to establish the functional relationships between each set of interacting components. Recalling the three requirements for a function to exist from Section 7.2.1, it is known that requirement one (there are two material objects) and requirement two (that interact with each other) have both been met. It is now simply necessary to keep requirement three (and the acting object changes or maintains a parameter of the recipient object) in mind to ensure that the functional relationships between the interacting objects are correctly captured. As an example, what is the functional relationship between the blacksmith and the hammer? It is understandable that the blacksmith must hold the hammer; therefore, "holds" is one of the functions that the

Table 7.3   Component Analysis of Blacksmith Operations

Blacksmith Operations	blacksmith	fire	metal rod	tongs	hammer	anvil	glove	air
blacksmith					Yes		Yes	
fire			Yes				Yes	Yes
metal rod		Yes		Yes	Yes	Yes		Yes
tongs			Yes				Yes	
hammer	Yes		Yes					
anvil			Yes					
glove	Yes	Yes		Yes				
air		Yes	Yes					

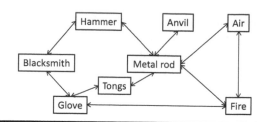

Figure 7.5   Metal-forming component interactions.

blacksmith performs on the hammer. But for the hammer to do something to the metal rod, then clearly something else must happen to the hammer for the hammer to create the work output. The blacksmith must make the hammer move (blacksmith moves hammer). Both actions by the blacksmith onto the hammer (holds and moves) result in the blacksmith changing or maintaining a parameter of the hammer (holds, maintains position of hammer in relation to hand and moves, changes the position of the hammer in 3D space). Considering a second functional relationship within the model, there is the function between the hammer and the metal rod. One representation could be hammer strikes rod. While this is true it does not necessarily result in a change or maintenance of one of the rod's parameters. Also, simply striking the metal rod is not the desired result. What does the hammer actually do to the metal rod? Why does the blacksmith strike the rod with the hammer? The reason a blacksmith strikes a metal piece with a hammer is to change the metal piece's shape. Therefore, hammer shapes metal rod is a better description and also results in changing a parameter of the metal rod, namely its shape. Figure 7.6 shows a completed model where the issue being identified is that the air is cooling the metal rod in a harmful manner (it is not desired for the metal rod to be cooled during the shaping operations). Other issues could also be represented in the model such as the metal rod shape not being changed as desired by the blacksmith. In this situation, it would be useful to show an insufficient relationship between the hammer and the metal rod. Further, the insufficiency of the hammer shaping the metal rod may very well be further related to the insufficiency of how the blacksmith moves the hammer. These insufficient relationships can be seen in Figure 7.7. The interesting thing about the chain of insufficiencies in this second model is that maybe the issue of hammer insufficiently shapes metal rod can be addressed in either location. Maybe the blacksmith can move the hammer differently, or maybe the hammer can be changed so that it better shapes the metal rod, or both.

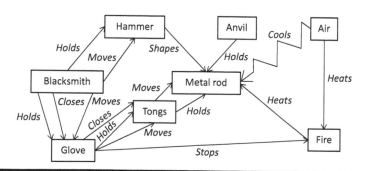

Figure 7.6   Metal-forming functional model 1.

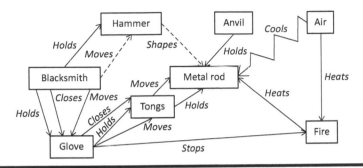

Figure 7.7   Metal-forming functional model 2.

7.6 Functional Analysis

7.6.1 Identifying Contradictions within the Model

Now that there is a functional model of the blacksmith's work, an analysis of the model can be executed. In Figure 7.6, it was established that air harmfully cooling the metal rod was the issue to be addressed. Admittedly, this is a distinctive situation because the only interface between the air and the metal rod is the harmful effect of cooling. In general, if the only relationship between two components is harmful then simply separate the components to stop the harmful relationship. Unfortunately, the component air is everywhere within the system, and therefore separation is difficult, if not impossible. Could the metal rod be worked within a vacuum chamber? If so, would the vacuum stop the unwanted heat transfer or rather make it worse? Regardless, the addition of a vacuum chamber that the blacksmith must work "through" seems like a challenging and less-than-ideal solution direction. Therefore, it is necessary to look for other solution concepts. In Chapter 5—Mapping Invisible Resources, we advised to look at what local and available resources are present in the system under analysis that might be used in support of a solution concept. What resources are available in the functional model that might help solve the problem? In studying the functional model (see Figure 7.6), it should be understood that any component in the model might be capable of contributing resources toward solving the problem of air harmfully cools metal rod. Since the problem is in relation to heat transfer then heat sources might be helpful, which of course leads to the fire. Further, as we can see from the functional model that the fire also heats the air. Therefore, fire, air, and their relationship might be useful resources in solving the problem of air cools metal rod.

To innovate a solution to this challenge, it would be helpful to model the problem at the next level. To do this, a tradeoff contradiction of the problematic relationship can be written.

- IF a heated metal piece is worked in an open area
- THEN the work space is somewhat uncomplicated
- BUT the ambient air will cool the piece too rapidly

As suggested previously, various principles can be chosen and applied toward a solution. In Chapter 3, it was introduced that selected 40 Principles could be identified by use of Altshuller's matrix in the pursuit of resolving contradictions. The matrix and the 40 Principles are considered in detail in Chapter 11. Principle 17—Another Dimension refers to moving things in three-dimensional space. Could the metal rod be moved closer to the heat source (fire)? Principle 13—"The

Other Way Around" refers to changing the way things are done and how to do them "the other way around." Instead of heating the metal rod in the fire and then working it outside of the fire, is it possible to heat the metal rod outside of the fire and then work it within the fire? What about heating within the fire and then also working it within the fire? Not exactly "the other way around" but close. Now, recalling the resource analysis of fire, air, and heat, and the principles Another Dimension and "The Other Way Around" triggers the solution concepts:

1. Use a fan and a metal duct to blow hot air from the fire onto the metal rod while it is being worked (resource utilization).
2. Heat and work the metal piece in the fire (resource utilization and "The Other Way Around" principle).
3. Heat the metal rod in the fire and then work it on an anvil that is placed just above the fire. (resource utilization and Another Dimension principle).

If the solutions are effective, then the problem is solved. If the solutions that were generated from the principles do not seem acceptable then other principles can be chosen randomly and also considered for solution generation. If the solutions are viewed as effective, but there appear to be challenges in implementation of the solutions, then a contradiction can be written concerning the implementation challenge(s), and once again the principles can be applied in solving the implementation issue(s). For particularly challenging implementation issues, the entire TRIZ process can be repeated for the new challenge.

As discussed, a functional model can have multiple challenges captured within it. Previously, a second issue of hammer insufficiently shapes metal rod was also identified (see Figure 7.7), which was serially linked to the function blacksmith insufficiently moves hammer. As previously detailed, the insufficient shaping of the metal rod might be addressed at either, or both, locations within the model.

1. Hammer insufficiently shapes metal rod
   a. IF a hammer is used to shape the metal rod
   b. THEN the tools are easy to operate
   c. BUT the shaping of the metal rod is difficult
   Apply principles to solve the problem:
   Principle 2—Separation
      Solution concept: Take out a portion of the hammer face to create a shaped hammer face
   Principle 5—Merging
      Solution concept: Merge a shaped hammer face with the standard blacksmith hammer to improve the metal-shaping process.
2. Blacksmith insufficiently moves hammer
   a. IF the blacksmith swings the hammer to strike the metal rod
   b. THEN the shape of the metal rod will be changed
   c. BUT the repeated accurate movement of the hammer is difficult
   Apply principles to solve the problem:
   Principle 10—Preliminary Action
      Solution concept: Utilize a shaped press to initially begin the formation of the metal rod
   Principle 40—Composite Materials

Solution concept: Use a composite metal hammer with different hardness properties purposely configured to affect the shaping of the rod
Principle 16—Partial or Excessive Actions
Solution concept: Use a large quantity of smaller, more controllable hammer swings to affect the shaping of the metal rod

It is important to remember that there is never only one correct solution path. Either, or both, contradictions could be written differently, and different principles could be considered toward the solution.

7.6.2 Increasing Ideality through System Trimming

Another use of functional analysis is to study and apply the concept of system trimming to increase the system's ideality. Trimming, as it sounds, is the elimination of system components, but in a way that preserves the necessary system functionality. If a component is removed from the system, then the complexity and cost of that system is reduced and the value of the system increases. Increasing the value of a system directly increases the system's ideality. As an example, if the frame of a pair of glasses is removed and the earpieces and nosepieces are assigned the function of holding the lenses, by way of directly connecting them to the lenses, then the ideality of the glasses has been increased. Taking this a step further, if the earpieces and nosepieces are removed from the frameless glasses and the support of the lens is now assigned to the eyeballs (contact lenses), the ideality of the system has once again been increased.

The old method of writing a check to pay for purchases has been replaced with debit cards. The movement of the physical check along with the people and equipment needed to receive, transfer, deposit, and transfer funds from account to account has been replaced with a completely electronic method. While computing and network infrastructure had to be developed to allow the trimming of paper checks from the "cash payment" system, that evolution removed a substantial number of physical and human components from the paper check system. As demonstrated, it is often the development of other systems that predicates and allows the component trimming of older systems.

The trimming opportunities of the metal rod–shaping system are limited as the system is already simple. However, there is at least one possible trimming scenario that could be pursued. Blacksmiths wear heat protective gloves when they work (for simplification purposes only one glove, with the functions of "holds," "closes" and "moves tongs," was included in the models [see Figures 7.6 and 7.7]). If the functions of holding and moving the metal rod are assigned to the blacksmith (made possible by the blacksmith's heat protective gloves), then the tongs can be removed from the system, thus increasing the overall system's ideality (see Figure 7.8).

While there is a lot of detail to the process of system trimming, which cannot be effectively covered in an introductory discussion, there are a few guidelines that simplify and clarify the process:

1. If the system has an expensive component, try to trim it.
2. If the system has a problematic component, try to trim it.
3. If you can reassign a component's functions (identified by arrows coming from that component) to another component, the original component can be removed.

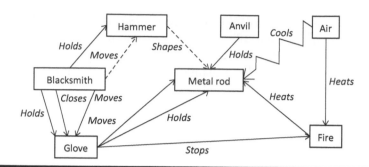

Figure 7.8   Metal-forming trimmed functional model.

4. Any functions originally affecting a trimmed component can be ignored and removed from the model. Once the function-affected component is removed, there is nothing to perform those functions on.

7.7 Technical and Business Case Studies

Now that the basics of functional modeling and functional analysis have been covered, it is time to start practicing the use of the tools. Like any skill set, functional modeling and functional analysis are best learned through repeated application. Though the methods might seem challenging at first, building a few functional models and identifying and solving contradictions within those models (functional analysis) is the best way to quickly come up to speed with the skill set. In moving toward that goal, it is helpful to see more models and to understand how they are used. One technical case study and one business case study showing functional modeling and analysis are presented in section 7.7.1.

7.7.1 Wireless Power System Improvement: Technical Functional Analysis Application

7.7.1.1 Introduction

In late 2010, the author was asked to support Intel Labs in identifying solution concepts around improving the effectiveness of a wireless power system.

7.7.1.2 Background

A wireless power system is an engineering system that transmits electrical power "through the air" as opposed to through a solid conductor, such as wiring. Not needing an electrical power outlet, or batteries, would be a big plus for a wide variety of electrical appliances including mobile computing platforms. The fundamental issue was that while the development team had a functional system, they were interested in improving its transmission efficiency. It is a fact of wireless power systems that the distance the power can be efficiently transmitted is directly proportional to the diameters of the sending and receiving power antennas (see Figure 7.9). The challenge was that the design was to place the transmitting antenna in the side panel of a laptop computer and the receiving antenna in the side panel of a mobile phone so that a user could charge her phone by simply laying it next to her laptop computer. This design effectively limited the diameters of the sending

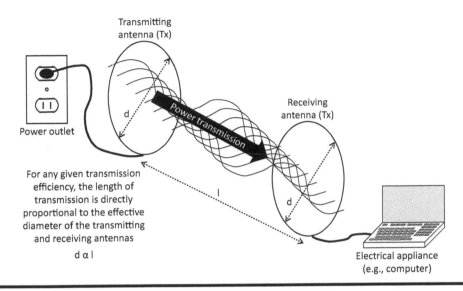

Transmitting
antenna (Tx)

Power outlet

Power transmission

Receiving
antenna (Tx)

d

d

l

For any given transmission
efficiency, the length of
transmission is directly
proportional to the effective
diameter of the transmitting
and receiving antennas

d α l

Electrical appliance
(e.g., computer)

Figure 7.9   Relation between antenna diameter and transmission distance for wireless power system.

and receiving power antennas to less than an inch. These small diameter antennas worked fine when they were placed very near to each other. Therefore, when the wireless power sending and receiving antennas were placed close together (almost touching), the resulting power transmission was good (high efficiency). However, the further the sending and receiving components were from each other, the worse the power transmission was (low efficiency). Consequently, if a user placed her phone several inches from the laptop, then the phone charging would take a very long time, if it charged at all. Improving the overall transmission efficiency would reduce the power usage of the system, increase the range over which it could operate (power transmitter to power receiver distance), or both. For this particular problem, there was already an operating system, but Intel Labs was interested in improving its performance and, therefore, improving the interaction between the system components.

7.7.1.3 Functional Modeling

Having an accurate functional model is crucial to the success of any project as an inaccurate model will lead down the wrong solution paths (garbage in, garbage out). To ensure an accurate functional model, it is best to engage system-level experts in the feedback and review of the model. Enter the Wireless Power Team (Intel project team) from Intel Labs. The same folks who needed some different ideas as how to solve a problem were the very folks who have the expertise to ensure that the initial modeling was accurate. Further, that team was also the perfect group to give feedback during the solution generation process. The use of the project "customer" as the system expert is an example of the utilization of locally available resources (see Chapter 5) in problem modeling and solution generation.

A block diagram of the wireless power system was requested (see Figure 7.10). As shown in Figure 7.10, State 1, when the transmitting and receiving coils are close together the system has good efficiency, and therefore the load current level is almost equal to the drive current level (good transmission efficiency). As shown in Figure 7.10, State 2, when the transmitting and receiving

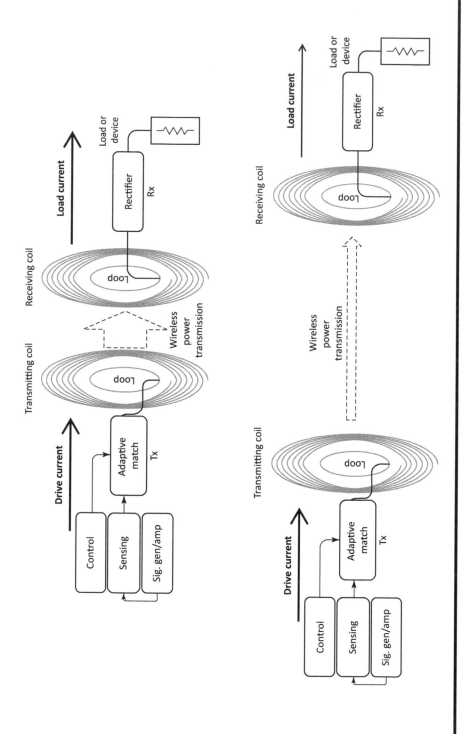

Figure 7.10   Wireless power system block diagram: (a) State 1 (good power transfer) and (b) State 2 (poor power transfer).

coils are far apart, the system has poor efficiency, and therefore the load current level is substantially smaller than the drive current level (poor transmission efficiency).

The diagram shows all the material (physical) components of the engineering system. However, not all the components necessary to create a complete functional model are shown in the diagram. Recall that per the TRIZ definitions, a component can be either a material object or a field (thermal, acoustic, electromagnetic, etc.). Therefore, the block diagram is missing some key components (field components) required to build a complete functional model. It should be noted that the transmitting (Tx) side of the system is made up of a physically disconnected driver loop and driver coil. Further, the receiving (Rx) end of the engineering system is also made up of a load loop and a load coil that also are not physically connected. And, of course, the Tx and Rx sides of the system are physically separated which represents the "wireless power" feature of the engineering system. The way a wireless power system works is by way of Faraday's Law. Faraday's Law says that when an electrical current is moved through a conductor, a corresponding magnetic field sweeps around that conductor. Accordingly, when a magnetic field is passed over a conductor, it induces a corresponding electrical current within that conductor. It is Faraday's Law that effectively drives the power transfer "device." Therefore, the electrical current and electromagnetic flux (EMF) are the missing components.

Now that the material components, and the existence of field components, have been identified, all the component categories needed to create a functional model are accounted for: material components (e.g., power supply, rectifier, transmission coil) and field components (e.g., current, electromagnetic flux).

One method to generate a functional model is to examine a diagram of the system under analysis and start building the model component by component. Referring to Figure 7.11, the first components (material and field) are documented. It should be noted that the control, sensing, and adaptive match components in the block diagram were combined into a single feedback and control component for the functional model simply because that truncation was sufficient for the analysis, as the issue was known to reside within the transmission and receiving components of the system. Had the issue been known to exist within the control, sensing, and adaptive match components of the block diagram, it is quite possible that the model would have been built differently with more focus on the interrelationship between those specific components. Also, it should be noted that the first field component, Current 1, was also included in the initial functional modeling. As shown, Current 1 is the electrical current produced by the power supply and controlled by the feedback and control component. Moving on to Figure 7.12, the next portion of the functional model was constructed. As we can see, there are two material components (driver loop and transmitting coil [Tx Coil]), and there are two field components (electromagnetic flux one [EMF 1] and Current 2). EMF 1 is the field produced by Current 1 as it is conducted by the driver loop, and Current 2 is the electron flow created by EMF 1 as it passes through the transmitting coil (Tx Coil). The reason that Current 1 and Current 2 (and the soon to be seen Current 3 and Current 4) are listed separately is because they are, indeed, separate currents performing separate functions on different components. As we will see in Figure 7.13, the same is true for the three different electromagnetic fluxes in the system: EMF 1 (Driver Loop and Tx Coil electromagnetic flux), EMF (electromagnetic flux between Tx Coil and Rx Coil), and EMF 2 (Load Loop and Rx Coil electromagnetic flux). Next, moving on to Figure 7.13, the receiving antenna was constructed as a mirror image of the transmitting antenna. The only differences are the component names and the fact that the electromagnetic flux (EMF) that carries the power from the transmitting to the receiving antenna is including in this portion of the functional model. Next, the last portion of the circuitry was added, as can be seen in Figure 7.14. Finally, three other components were added to

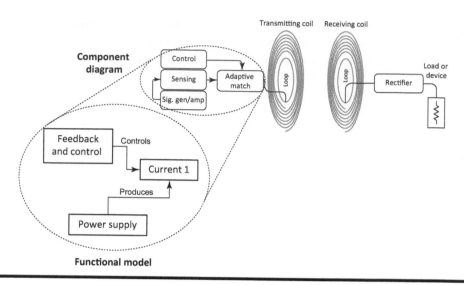

Figure 7.11 **Wireless power system functional model—transmitting circuitry.**

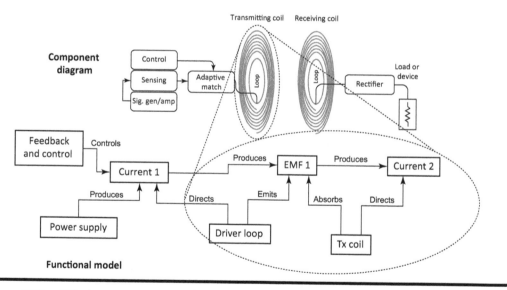

Figure 7.12 **Wireless power system functional model—transmitting antenna.**

the functional model (see Figure 7.15). These components do not appear in the block diagram of the wireless power system but were considered because of their possible contribution to the problem and the solution. Gravity was included as a component because of its effect on the antenna's orientations and the conductors and insulators components were added because of their known effects on EMF. These last three components are considered supersystem components because they are not part of the system under analysis but do interact with the system under analysis in a significant way. As an example, if a functional model of a car cover is being built to study how its protection of the car can be improved, it would be necessary to include environmental supersystem (macrosystem) type components such as wind, rain, sunshine, and dirt.

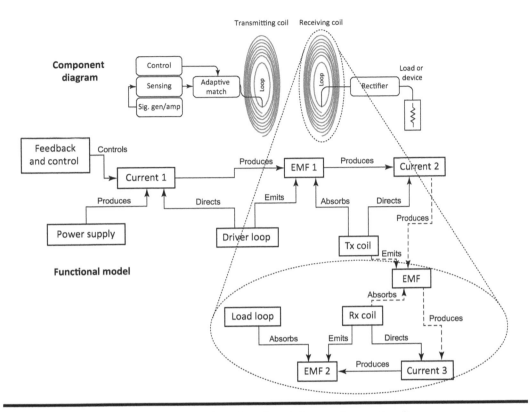

Figure 7.13   Wireless power system functional model—receiver antenna.

7.7.1.4 Functional Analysis

Now that the wireless power system functional model has been built, how are problem areas identified to create a drill-down problem model? Remember that there are two general categories of functions: useful and harmful. Further, there are three subcategories of useful functions: sufficient, insufficient, and excessive. It is logical to assume that useful–sufficient functions are just that, useful, and sufficient, and therefore do not need to be improved. That leaves harmful, useful–excessive, and useful–insufficient functions. All the harmful, useful-excessive and useful-insufficient functional relationships between components are good candidates for contradiction modeling. Examination of the functional models shows that there are harmful functions identified between two of the supersystem components (conductors and insulators) and the electrical transmission (EMF). Further, there are insufficient functions identified between four components (Current 2, Tx Coil, Rx Coil, and Current 3) and the electrical transmission (EMF). For this problem, the Intel project team agreed that the harmful interactions shown could be ignored and instead concentrated on the insufficient interactions. While the functional relationship of Current 2 insufficiently produces EMF is an issue, it does so because of the small size of the Tx Coil (see Figure 7.15). It was, therefore, prudent to build a contradiction analysis around the small size of the transmitting (Tx) and receiving (Rx) coils in the system:

- IF small Tx and Rx coils are used in the wireless power system
- THEN the coils will fit in the side panels of the computer and mobile phone
- BUT the transmission distance of the wireless power will be very small

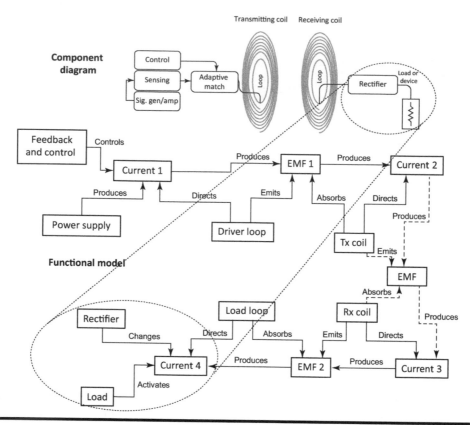

Figure 7.14   Wireless power system functional model—receiver circuitry.

Apply principles to solve the problem:

- Principle 17—Dimensionality Change
  Solution Concept: One solution concept was to change the dimension of the coils or even move (in three-dimensional space) where the coils were located, or both. The author proposed putting the transmitting coil in the lid of the laptop (greatly increasing its diameter and, therefore, creating an efficient transmission distance). However, this left the issue of the receiving coil size and location. One option developed was that the receiving coil could be built into a phone case and could be hinged and gimbaled so that it could be folded out or flipped up. This solution did improve the performance of the system but required some work by the user in positioning the receiving antenna—a less-than-ideal solution direction.
- Principle 32—Optical Property Changes (Color Change)
  Solution Concept: Referring to Chapter 11 it is explained that color change can also be used to trigger thoughts of frequency change of light/electromagnetic fields (EMF). As it turns out, lower frequencies of EMF performed better than higher frequency EMF when moving through air due to less absorption by the air. Further, higher EMF frequencies require more power input than lower EMF frequencies. Principle 32 drove the project team to reconsider the transmission frequency, and the system performance was somewhat improved by lowering the transmission frequency of the wireless power system.

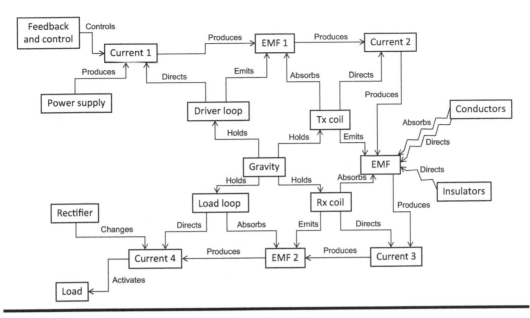

Figure 7.15   Wireless power system functional model.

While the results of the contradiction analysis did improve the wireless power system performance somewhat, the Wireless Power Team desired even better performance. Therefore, for the original project executed in 2010, other TRIZ solution development methodologies were also utilized in addition to the application of the 40 Principles. More specifically, the wireless power solution directions were also explored with application of the patterns of system evolution and the application of resource analysis guided by MATChEM (see Chapter 10—Patterns are Powerful Tools for System Development).

7.7.2 Wireless Power System Improvement: Patterns of System Evolution Application

The patterns of system evolution were introduced in Chapter 2 and are discussed in more detail in Chapter 10. The application of one of the patterns toward the solution for the wireless power problem can be found as an example in Section 10.5.1.

7.7.3 Wireless Power System Improvement: Application of MATChEM

In Section 5.5.7, the analysis of the energy fields of mechanical, acoustic, thermal, chemical, and electromagnetic (MATChEM) was discussed. The method involves considering the use of different, or multiple, energy fields when looking for resources by which to innovate a solution to a problem. The initially analyzed wireless system utilized an EMF to transmit the "electrical power." The MATChEM methodology suggests considering different, or combinations of, fields to solve the problem. An additional solution concept could be generated by considering how to use one field to cause the creation of another field. This prompted the consideration of different types of energy fields arriving at the following somewhat more complex, but ultimately workable, solution. Instead of directly transmitting power, use the Tx Coil current to trigger a sound generator that sends ultrasonic sound waves (outside of the human hearing range) that are received by

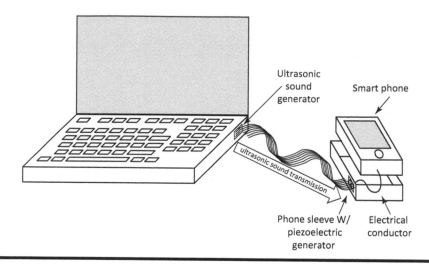

Figure 7.16   **Ultrasonic sound transmission with piezoelectric current generator receiver.**

a device that generates electricity when triggered by those sound waves. The final solution was to utilize a piezoelectric nano-wire energy generation system built into a phone sleeve that picks up the sound waves and generates power for the cell phone, as opposed to transmitting the power over a distance. In this manner, the somewhat limited range of the transmitted EMF was effectively replaced by transmitting sound waves (acoustical) and then using those sound waves to drive the generation of electrical current directly where it was needed (see Figure 7.16). Therefore, instead of transmitting EMF, electrical current is instead created at the point of need.

7.7.4 Call Center Improvement: Business Functional Analysis Application

7.7.4.1 Introduction

The application of TRIZ to technical problems is aimed at solving technical challenges (e.g., how to improve the performance of a machine, how to overcome manufacturing challenges) and, therefore, most often looks at technical systems at a subsystem interaction level (e.g., How does a pulley interact with the belt that is turning it?). In contrast, the application of TRIZ to business problems is aimed at solving business challenges (e.g., How to improve customer service, how to better utilize equipment within a business environment) and, therefore, most often looks at business systems at a system-to-system interaction level and is more likely to included human "systems" as operational entities (e.g., How can a salesperson better influence potential customers?). While an analysis often uncovers both technical and business issues, most analyses are usually initiated with either a technical or business pursuit in mind.

7.7.4.2 Background

In 2014, the author was visiting a client in India and was asked to help analyze and improve the technical support provided by their call center. In today's world of high technical devices, live technical support is becoming more and more of a challenge. In fact, the challenge of properly training sufficient numbers of technical support personnel is difficult enough that many companies have simply created online technical information access where the customers fundamentally research and solve their own problem, or not. As many have experienced, online technical information access systems often leave customers feeling poorly supported and frustrated when trying to understand how best to use their new products.

These two service models (live technical support and online technical information access) have opposite sets of pluses and minuses that can be reflected as tradeoff (or inherent) contradictions.

Live technical support:

- IF live technical support is utilized to provide technical product support to the customer
- THEN the technical support operator can easily react to customer's different needs
- BUT the breadth of knowledge required of the technical support operator is wider than can be absorbed in a reasonable training time period

Online technical information access:

- IF online technical information access is utilized to provide technical product support to the customer
- THEN a very broad base of technical information can be made available to the customer
- BUT easily accessing the correct technical information can be very challenging for a customer who does not know the new product well in the first place

7.7.4.3 Functional Modeling

In this situation, it is possible to functionally model one of the two systems (live technical support or online technical information access), and then use the positive attributes of the other system to help establish system improvement goals. It is always best to model a system as it currently exists in order to capture the current challenges with the system and to document the locally available resources for use in solution generation. In this case, the existing system was the live technical support center. The following shows the development of a live technical support system functional model (problem model). It is important to remember that when building functional models, the problem solver either needs to know how the system currently works, or they need access to experts who can provide that information. In this manner, it is fairly easy to correctly capture the components that comprise the system.

As previously discussed, it is best to first create an interaction table of the components in the system (see Table 7.4). It should be noted that the system component "air" was included in the component listing to show the inclusion of a supersystem component within this analysis. It is quite possible that the problem could have just as easily been solved without the inclusion of that particular system component.

There are other assumptions captured in the component listing that are important to discuss. First, the live technical support is assumed to be phone based but could just as easily be reflected as an online live support. Second, only one phone is captured in the component listing because the two phones (customer's and agent's) operate in the same way and provide the same functionality. Further, the challenge at hand is not concerned with the effectiveness of the telephone system. In other words, the phone system is assumed to work fine, and therefore there is no need to show both telephones, the switching equipment, or the phone lines. If the technical operation of the phone system was the primary concern, then the system would be modeled at a different level and would more than likely become more of a technical analysis than a business analysis. Next, it may

Table 7.4   Component Analysis of Live Technical Support

Live technical support	policies	supervisor	customer	telephone	agent	computer	procedures	headset	air
policies		Yes			Yes				
supervisor	Yes				Yes		Yes		
customer				Yes					
telephone			Yes			Yes		Yes	Yes
agent	Yes	Yes				Yes		Yes	Yes
computer				Yes	Yes		Yes		Yes
procedures		Yes				Yes			
headset				Yes	Yes				
air				Yes	Yes	Yes			

be noticed that the two components of procedures and policies are included in the listing. Some readers may question whether these components are really material objects. Before the dawn of computing, most would agree that these components would certainly be material objects as they would have been hardcopy documents recorded on paper and stored in a file cabinet. Today, these components would be a collection of electronic charges stored on a computer memory (both material objects) and structured by way of a register. While this discussion is beginning to crossover with TRIZ for computing topics, it should be understood that document components should be considered as material objects and, therefore, treated as components, within functional modeling of business systems.

Once the interaction matrix was created, the next step was to create a map of the interacting components (see Figure 7.17). It should be recalled that a graphic of the component interactions simply shows which components interact with each other. It does not show what those interactions are, how many interactions there are, or the quality of the interactions.

When converting the interaction graphics to a functional model, it is necessary to assign functional relationships between the components and what the quality of those functions are (i.e., useful–sufficient, useful–insufficient, useful–excessive, and harmful). For the live technical support case study, it was assumed that the information given to the customer was insufficient. Therefore, it is necessary to establish an insufficient informs relationship between the agent and the customer. There are two primary reasons why the agent provided insufficient information to the customer: 1.) the customer could not describe the problem to the agent very well, and 2.) the support procedures were not written very well (i.e., the support procedures did not communicate high-quality problem resolutions steps to the agent). Referring to Figure 7.18, challenge number one (customer cannot describe the problem) was reflected as the following:

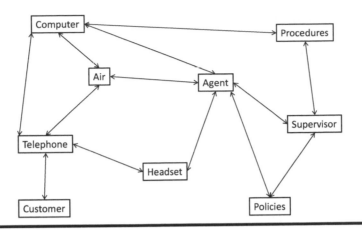

Figure 7.17   Live technical support component interactions.

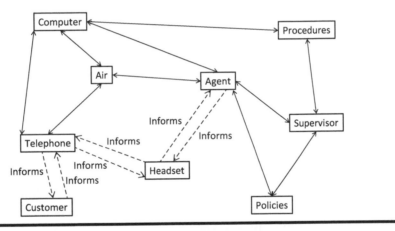

Figure 7.18   Live technical support functional model building—step 1.

Customer insufficiently informs the telephone
    Therefore
Telephone insufficiently informs the headset
    Therefore
Headset insufficiently informs the agent

(Note—In this scenario the inclusion of the component headset may not have been necessary and a functional model without the headset may have been just fine. In general, if a component's contribution to either the problem model or solution generation is unclear, it is best to go ahead and include that component in the analysis, even if in hindsight that component's inclusion was not necessary. It is better to overcomplicate a model with unnecessary components than to exclude components crucial to the analysis.)

Since the customer insufficiently describes the problem to the agent, it stands to reason that the agent insufficiently informs the customer of the problem resolution procedures:

Agent insufficiently informs the headset
    Therefore
Headset insufficiently informs the telephone
    Therefore
Telephone insufficiently informs the customer

Referring to Figure 7.19, challenge number two (support procedures are not written very well) can be reflected as the following:

Supervisor insufficiently creates procedures
(Since it was the supervisor who was responsible for the creation of the procedures)
    Therefore
Procedures insufficiently informs computer
(Since it was a computer where the procedures were stored)
    Therefore
Computer insufficiently informs agent
(Since the agent received the problem resolution instruction from the computer)

In reference to Figure 7.20, the remainder of the functional model was completed (left to right) by the following:
Capturing that there was a two-way control function between the computer and the telephone:

Computer sufficiently controls telephone
    And
Telephone sufficiently controls computer.

Capturing that the air damaged the computer because the computer heated the air (note: A business functional model can address both business and technical issues if need be. The inclusion of the air in this model was in relation to solving a technical issue of the computer and agent overheating in the crowded and busy call center.):

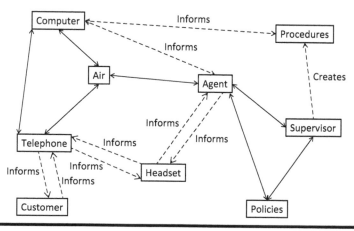

Figure 7.19   Live technical support functional model building—step 2.

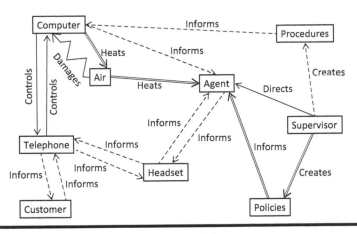

Figure 7.20   Live technical support functional model.

Air harmfully damages computer
        Because
Computer excessively heats air
        And
Air excessively heats the agent

 Capturing that the polices were excessive and overcontrolling:

Policies excessively informs agent
        Because
Supervisor excessively creates policies

 Capturing that the supervision of the agent was just at the level needed:

Supervisor sufficiently directs agent.

7.7.4.4 Functional Analysis

With the functional model completed all, or any, of the insufficient, excessive, or harmful functions could be addressed by way of a contradiction analysis. Focusing on the existence of poor customer support procedures (which may also help the agent in supporting a customer who has difficulty describing their problem), the following tradeoff contradiction was written.

- IF the supervisor creates the customer support procedures
- THEN there is efficiency in procedure creation for all the agents
- BUT the supervisor may not be familiar enough with all the potential customer issues to create comprehensive support procedures

 Apply principles to solve the problem:

In applying the principles to business problems, it is important to remember that oftentimes analogies of the principle's descriptions should be used.

- Principle 24—Intermediary
  Solution Concept: Use an intermediary phone menu to help the customer focus in on their issue, and then direct the call to the correct customer service agent. Phone menus directing customers to the correct customer service agent have been in use for some time, but the concept of having those phone menus provide a clearer understanding of the problem to the customer (by way of asking questions to help the customer realize a more precise description of their problem) is not common.
- Principle 26—Copying
  Solution Concept: Use a graphical interactive copy of the product on the technical support computing system that the customer can direct the agent through to discover the same issue that the customer is not able to sufficiently describe.

7.7.5 Call Center Improvement: Deep System Trimming

In Section 6.2, it was discussed that one of the ways to improve a system's value is to reduce the cost of that system. One of the easiest ways to reduce the cost of a system is to eliminate components from the system while maintaining the system's main functionality. Removing components while maintaining functionality is called "system trimming." Removing a significant number of components, while maintaining system functionality, is called "deep system trimming." This section will show an example of deep system trimming as applied to the original India call center case study. It is important to keep in mind that the deeper a system is trimmed the more radical the ultimate changes to the system will be and the greater chance of substantially increasing the system's ideality.

When trimming a system, the goal is to remove problematic and expensive components while maintaining or improving the system's functionality. Referring to Figure 7.20, there are multiple insufficient relationships, four excessive relationships, and one harmful relationship. When trimming systems, a trimming model is created which is a functional model of the trimmed system without necessarily having the insight as to how the trimmed model is to be executed in reality. In other words, the trimming exercise is just that, an exercise to remove components and reassign their functionality, without necessarily knowing exactly how the system might operate without the trimmed component. With that in mind, the deep system trimming was started by removing an expensive and problematic component: the agent (see Figure 7.21). Once a component is removed, all functions affecting that component (function arrows point toward the trimmed component) can be eliminated as those functions are no longer necessary. All functions created by the trimmed component (arrows pointing away from the trimmed component) must either be reassigned to another component or the component they lead to must also be trimmed. Therefore, it was then necessary to remove the headset so that the function "informs" that came from the trimmed agent was no longer necessary (see Figure 7.22). Next, since there is no longer an agent, there is no reason to have policies and if another way can be found to create the procedures then there is no reason to have any supervisors (see Figure 7.23). Since the air causes damage to the computer it was also wise to remove the air (which is no longer an issue for the humans as there are not any left in the service center) (see Figure 7.24). After those trimming steps, it was necessary to reassign the "hanging" functions of "creates" and "informs" to other components. Further, a new function of "informs" was added to allow the telephone to converse with the computer (see

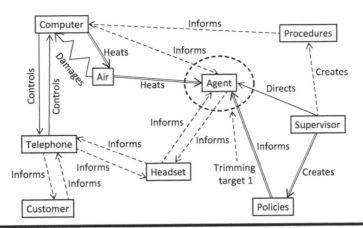

Figure 7.21   Live technical support functional model—deep trimming.

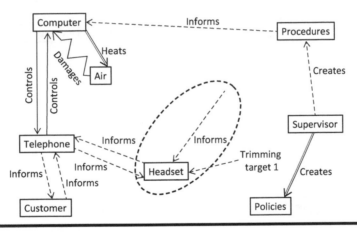

Figure 7.22   Live technical support functional model—deep trimming step 1.

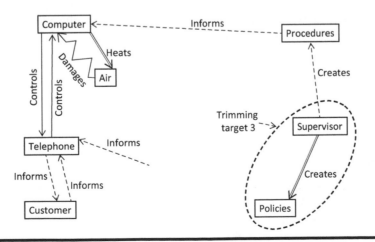

Figure 7.23   Live technical support functional model—deep trimming step 2.

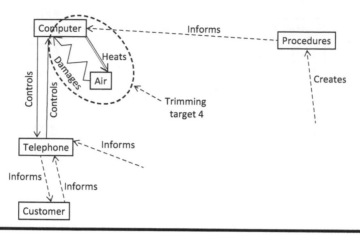

Figure 7.24 **Live technical support functional model—deep trimming step 3.**

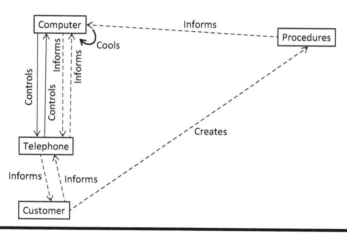

Figure 7.25 **Live technical support functional model—deep trimming step 4.**

Figure 7.25). The final call center design was built around customer-driven procedures built automatically by the computer by way of customer experiences that streamlined the process by capturing the most prevalent customer issues and the solutions based on the customer's direct experience using the product in question. This eliminated a system based on an agent giving advice regarding a product he or she may, or may not, have ever used and based on procedures written by a supervisor who also may not have much experience with the product. Further, all air-cooled computers were replaced with water-cooled systems.

7.8 Summary

- Functional analysis highlights how a system works and focuses the problem solver on the most important and challenging aspects of those workings.
- List the components involved in the operation of the system and include macrolevel (supersystem) components (e.g., air) when necessary to properly describe the system and its challenges.

- Utilizing an interaction matrix, understand which system components (including super-system) have direct physical contact with each other. Those components will have function relationships between them.
- Draw a map of the interacting components and define their functions, directions, and categories (useful, insufficient, excessive, and harmful).
- Any pair of components that have an insufficient, excessive, or harmful interaction (function) between them is a candidate for contradiction modeling (see Chapter 3—Clarify the Tradeoff behind a Problem and Chapter 4—Moving from Tradeoff to Inherent Contradiction) and solution generation.
- A system's ideality can be increased by removing (trimming) some of its components. If a component is removed from its associated functional model, all arrows (functions) leading to it can be eliminated from the model and all arrows (functions) leading away from it must be reassigned to other components, or the need for those functions must be eliminated.

References

1. Libraries, University of Wisconsin-Madison. 2017. The Lawrence D. Miles Value Engineering Reference Center collection, January 10. https://minds.wisconsin.edu/handle/1793/301.
2. SkyMark Corporation. 2016. Larry Miles and value engineering. http://www.skymark.com/resources/leaders/larrymiles.asp.
3. Johnson, P. 2013, July 10th. Fast diagramming made easy: Straightforward techniques for your highway project, Slide 5. Presented at the 2013 AASHTO Value Engineering Peer Exchange Workshop, Minneapolis, MN. http://design.transportation.org/Documents/TC%20Value%20 Engineering/2013%20VE%20Workshop/2013%20PPPs%20Papers_Wednes%20AM/1-Fast%20 Diagramming%20Made%20Easy.pdf.
4. Litvin, S., Feygenson, N., and Feygenson, O. 2011. Advanced function approach. *Procedia Engineering*, 9, 92–102. http://www.sciencedirect.com/science/article/pii/S1877705811001202.

Chapter 8

How to Separate the Best from the Rest: A Simple and Effective Tool for Evaluation of Solutions

8.1 Introduction

Early in the book, we asked you to recall your best problem-solving experience and think about what characterized the good ideas. This chapter begins with another question. When you create a good idea, do you ever wonder: "Why not until now? Why didn't I think of this 2 years, 5 years, 10 years ago?" Companies tell us this story so often that we have named it the "Standard Story"—"A competitor introduced a new solution and we found the same idea in our own notes from many years ago." Chapter 1 has many examples of good ideas that were neglected.

One of the most striking results from the authors' experience in teaching creativity classes and consulting on creativity is that recognizing, appreciating, and evaluating solutions may be more difficult than finding them. Having good ideas is useless if they are rejected.

We hope you agree that it makes sense to seek better ways to evaluate solutions. In this chapter, we will present a simple and effective evaluation tool considering the following three points:

1. We define the evaluation criteria, which we obtain from the concepts of ideality, contradiction, and resources (covered in Chapters 3 through 6). In this chapter, we will use these tools for a new purpose: the evaluation of proposed solutions.
2. We consider the measures of evaluation. The ideality of each proposed solution is evaluated and compared with the ideality of other solutions. The assessments are made by way of a simple and practical method, a pairwise comparison with other known solutions. In real-life projects, the standard should be the best possible existing or developing competing methods or technologies. For clarity and simplicity, our example solutions are usually compared with well-known current technologies.

3. We discuss how to go further if the evaluation shows that we have not gone far enough in the quest for the Ideal Final Result. Sometimes, ideas are bad and deserve to be rejected. More often though, the primary idea is excellent, but there are subproblems that need to be resolved. The evaluation criteria will help you see the path through the maze of problems and solutions and better understand the numerous secondary tasks requirements.

8.2 Evaluation Criteria

When we have a new idea, we must ask, "Is this idea good or bad?" To answer the question, we need a set of criteria for good solutions. Let us recall the ideality equation studied in the previous chapters:

$$Ideality = \Sigma Benefits / (\Sigma Costs + \Sigma Harm)$$

Ideality is the basis for evaluation.

- First, all harmful features disappear. Most often, problems are addressed because of some drawback, which is why it is logical to begin from this requirement. Removal of all harmful features may not be possible, but the goal is to drive as far down that goal vector as possible.
- Second, all useful features are retained and new benefits appear. We do not—and should not—remove only drawbacks. New useful features should be introduced when possible and existing ones retained.
- Third, new harmful features do not appear. It is important to check this—a frequent problem with both business and product improvements is that the new system gets rid of the initial problem, but introduces new problems. The software industry is legendary for "improvements" that cause customer dissatisfaction.
- Fourth, the system does not get more complex (complexity increases cost and reduces reliability).
- Fifth, the solution removes the inherent, primary, most important contradiction in the problem. Having studied tradeoffs and inherent contradictions in Chapters 3 and 4, we can see this requirement in the ideality equation, as well. To enjoy benefits, we need more weight, size, energy, time, and other cost-generating features. To cut cost and avoid harm, we should have less weight, size, energy, time, and other properties; always ask yourself "what is the essential primary contradiction?" The answer is the primary problem that should be solved. Before the car was developed, vehicles used steam engines. The basic contradiction of the steam-powered vehicle was the relationship of power to weight. The more power, the more weight. The engine should be heavy (to produce enough energy), and the engine should be light (to be manageable on the road without rails). In the late 1800s, the internal combustion engine resolved the contradiction of the steam engine rather well. The electric car, of the time, did not because the fuel quantity (battery size) required to move the electric car a required distance was so large that the power-to-weight ratio did not support the system's evolution away from the steam engine. That is why the car with the internal combustion engine won, although the electric car has many other benefits. Now, more than a century later, we are seeing the development of new kinds of electric cars that may finally replace internal combustion automobiles, mostly because of advancements in the power densities of the latest technology electric batteries. However, even if battery power densities

never outpace those of fossil fuels, a different technology jump might propel the electrical vehicle to a position of higher ideality than its fossil fuel competitor. Inductive coil electrified roadways or even satellite-based power beaming systems, both representing the trend of transition to the macrolevel (supersystem), could make the electric vehicle our primary transportation source in the somewhat near future.

■ Sixth, idle, easily available, but previously ignored, resources are used. Resource mapping was studied in Chapter 5. We can also find this requirement in the ideality equation. Benefits can be increased while cost is decreased if some new and inexpensive resources can be found.

These six evaluation criteria are generic, based on the fundamental concepts of TRIZ. Other criteria are specific to the particular system that is being studied, such as safety, speed of implementation, compatibility with existing systems, compliance with regulatory requirements (which may be different in different countries), or other issues. It is convenient to reserve a place for these miscellaneous criteria. Let us add the seventh criterion—other requirements.

Here are the seven criteria:

1. All harmful features vanish or are minimized.
2. All useful features are retained and new benefits appear.
3. New harmful features do not appear.
4. The system does not get more complex.
5. The primary tradeoffs and contradictions are removed.
6. Idle, easily available, but previously ignored, resources are used.
7. Other requirements related to the developed system are fulfilled.

8.3 Measures of Evaluation

Cost is not included explicitly in the list of criteria. We have found that if the idea is a real breakthrough, people will find ways to eliminate cost as a barrier. Further, money is simply a human construct to allow the easy comparison of things. You do not "reduce" money from a system to reduce its cost, you reduce the reasons for the expenditures of the money (rare materials, excessive human labor, complex manufacturing systems, etc.). In the same way, cost should never be listed in a contradiction statement; rather, the reason for the cost should be captured. TRIZ is used repeatedly—first solve the initial problem, then solve the focused problems of reducing cost, if need be. In one recent TRIZ class, people found a way to make a new product for cooking and selling individual portions of food. However, their management rejected the idea because the new factory required would be too expensive, based on the estimate of how much the product would cost and how much they could sell. The TRIZ class was not discouraged. They started a new project that reduced the cost of the proposed factory by 60%. This was enough to persuade their management that the new product could be a success.

Reading this, you may ask how this team was able to reduce the cost of the proposed factory by 60%. This was not the result of one big breakthrough, but rather the result of repeated applications of TRIZ to each of the processes in the proposed factory, focusing on improving the efficiency of each process. In this case, as in many others, we cannot publish the specific examples of the outstanding results achieved through TRIZ. A good solution is, by definition, a solution that gives the company a competitive edge. That is why companies such as Procter & Gamble and Ford, both of which have many years' experience with TRIZ, have published very few examples.

In using the list, you will find that comparison of pairs is very clear and much more reliable than attempting to define some absolute level of comparison. New and old technology should also be compared in the same time interval and environment. The ideal lawnmower today is different from the best method of controlling a lawn 5 or 10 years from now, which might be "smart grass" that keeps itself at the right height, rather than a grass-cutting machine. However, both concepts can be compared (in the current environment) to understand which is a more ideal system. The speed at which new concepts will be developed is often underestimated. Similarly, ideas for decreasing noise and pollution will undoubtedly get much more valuable as time goes by.

Discussion is meaningless if the circumstances for the use of the technology are not defined. There is a lot of discussion, for example, of energy technologies in general. Which is better, solar energy, nuclear energy, coal energy, hydropower, or something else? An answer is not possible before establishing for what purpose, location, time frame, etc., the energy is needed. Is it for an industrial plant consuming hundreds of megawatts around the clock, for a cottage using some tens of kilowatts temporarily, or for a space exploration vehicle en route to Saturn?

Using the criteria of TRIZ decreases the subjectivity of the evaluation, although it does not remove it totally. We do not claim that the set of criteria is 100% comprehensive or that it will mechanically produce an unambiguous result. We stress that it is most fruitful for the development team to discuss the evaluation criteria first and then make the evaluation.

8.4 Examples of Evaluation

The evaluation of new solutions can be easily presented in a table having two columns:

1. Presentation of the criteria, independent of the result (left column).
2. Evaluation (right column).

Let us evaluate some solutions presented in earlier chapters. As we have noted earlier, in these training examples, the measures for comparison are well-known current technologies. Examples are simplified to make the tool for evaluation as easy to use as possible. Using grass for noise suppression is compared with a big conventional muffler (Table 8.1). The seeding tape is compared with a precision seeder, a simple mechanism with small wheels (Table 8.2). In the case of the latching mechanism, the conical pin is compared with the cylindrical one (Table 8.3). Training embedded in work is compared with traditional classroom training (Table 8.4). Fighting fire with water mist is benchmarked against the use of drops in typical sprinklers (Table 8.5).

We do not try to claim that the examples present the best possible technologies. Actually, because the examples are proven and published solutions, they are inevitably at least somewhat out of date. However, we are sure this is not a problem. It is great if you have better solutions in mind. Simply insert your knowledge into the tables for comparison and tailor the examples for yourself. This book provides you with the best possible tools for the generation and evaluation of solutions. You will produce and select the best solutions yourself.

For your own examples, first describe the problem, the TRIZ solution, and the best conventional solution:

■ My problem
■ My TRIZ solution
■ Best conventional solutions

Then fill in the evaluation table (Table 8.6).

Table 8.1   Using Grass and Conventional Muffler to Suppress Lawnmower Noise

	Criteria	*Comparison with Known Solution*
1.	Do the harmful features disappear?	Yes. Noise will be decreased.
2.	Are the useful features retained? Will new benefits appear?	Yes. It cuts grass as well as the conventional machine.
3.	Will new harmful features appear?	No.
4.	Does the system become more complex?	No. It gets simpler.
5.	Is the inherent primary contradiction resolved?	Yes. Conflict of big muffler–small muffler is solved.
6.	Are idle, easily available, earlier ignored resources used?	Yes. Grass and geometry used.
7.	Other criteria: Easy to implement	Yes.

Table 8.2   Carrot Cultivation—Seeding Tape Compared with Precision Seeder

	Criteria	*Comparison with Known Solution*
1.	Do the harmful features disappear?	Yes. System gets simpler; no new devices are needed.
2.	Are the useful features retained? Will new benefits appear?	Yes. Accuracy is retained; speed is improved.
3.	Will new harmful features appear?	No. No new harmful features.
4.	Does the system become more complex?	No. It gets simpler. No need to buy or rent a machine.
5.	Is the inherent, primary contradiction resolved?	Yes. Conflict of one seed–many seeds is solved.
6.	Are idle, easily available, earlier ignored resources used?	Yes. Strips seed themselves; utilizes easily available materials.
7.	Other criteria: Easy to implement, usable in all size gardens?	Yes.

Table 8.3   Improving Latching Mechanism—Conical Pin Compared with Cylindrical Pin

	Criteria	Comparison with Known Solution
1.	Do the harmful features disappear?	Yes. Yes, there will be no wear.
2.	Are the useful features retained? Will new benefits appear?	Yes. Simplicity retained; the locking problem is solved.
3.	Will new harmful features appear?	No.
4.	Does the system become more complex?	No. It gets simpler.
5.	Is the inherent, primary contradiction resolved?	Yes. Conflict of big clearance–no clearance is solved.
6.	Are idle, easily available, earlier ignored resources used? Is the solution new?	Yes. The geometry of the pin is used.
7.	Other criteria: Easy to implement?	Yes.

Table 8.4   Training Incorporated into Work Compared with Traditional Classroom Training

	Criteria	Comparison with Known Solution
1.	Do the harmful features disappear?	Yes. Time is decreased.
2.	Are the useful features retained? Will new benefits appear?	Yes. Training results retained and even improved.
3.	Will new harmful features appear?	No.
4.	Does the system become more complex?	No (and a little yes). Good planning makes learning simple for trainees. Some complex work by the training designers is required to set up the system and measure its effectiveness.
5.	Is the inherent, primary contradiction resolved?	Yes. Contradiction lots of time–no time resolved.
6.	Are idle, easily available, earlier ignored resources used?	Yes. Working time is used.
7.	Other criteria:	

Table 8.5   Firefighting—Water Mist Compared with Droplets

	Criteria	*Comparison with Known Solution*
1.	Do the harmful features disappear?	Yes. Water consumption and damage decreased.
2.	Are the useful features retained? Will new benefits appear?	Yes. Capacity to extinguish fire retained.
3.	Will new harmful features appear?	Yes. No new harmful features.
4.	Does the system become more complex?	No and yes. Machinery for getting high pressure needed.
5.	Is the inherent, primary contradiction resolved?	Yes. Much water–no water solved nearly ideally.
6.	Are idle, easily available, earlier ignored resources used?	Yes. Water is used.
7.	Other criteria: Easy to train firefighters to use?	Yes.

Table 8.6   Evaluation Table

	Criteria	*Comparison with Known Solution*
1.	Do the harmful features disappear?	
2.	Are the useful features retained? Will new benefits appear?	
3.	Will new harmful features appear?	
4.	Does the system become more complex?	
5.	Is the inherent, primary contradiction resolved?	
6.	Are idle, easily available, earlier ignored resources used?	
7.	Other criteria:	

8.5 Improvement of the Solution

If your solution is nearly an Ideal Final Result, you should get five "yes" answers (1, 2, 5, 6, 7) and two "no" answers (3, 4) within your ideality evaluation table. However, your first idea—even the best basic idea—almost always contains drawbacks that should be removed to get a working solution. After the first formulation of the Ideal Final Result, we nearly always find new contradictions

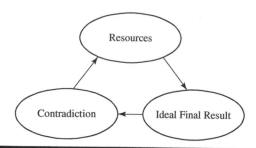

Figure 8.1   The improvement cycle. Repeating the process speeds the development project as a whole and improves the result.

and start again on the process of mapping resources to resolve them. The basic concepts of contradiction, resources, and ideality form a loop that is repeated many times to develop the new solution (see Figure 8.1).

Using grass as a muffler reduces lawnmower noise considerably. Certainly, this solution should be developed further, toward a totally noiseless lawnmower, a lawnmower without toxic exhaust gases, and eventually to the nonexistent lawnmower, the garden system that always keeps the grass the proper height.

How about a tape for seeding? There may be seeds in the tape that will not grow. It is necessary to fix seeds on the tape. The tape is a new component. These are new problems whose solutions are necessary to go forward from a prototype stage.

Water-jet cutting, developed in the 1960s, was slowly introduced in industry. Materials could be cut without any wearing of tools or generation of excess heat. However, these benefits were coupled with an annoying drawback. Slowly moving water is soft. The water must move at high speed to make it act as hard as an abrasive cutter. High-velocity water requires high pressure. To cut thick, tough materials, pressures of 2000–3000 bars are needed. So, we have a tradeoff between improved cutting properties and substantially increased complexity of equipment. Suppose we use the water to move microscopic cutting particles such as grains of sand? Then, the particles do the cutting and much less water pressure is needed. Indeed, about 20 years after the concept was first introduced, abrasive additives were brought in. Jet cutting is now in wide use.

There are many situations in which a rigid system is improved by adding hinges. The positioning of a lamp has been made more controllable through the use of flexible joints. The penalty has been more parts. The solution has been further improved by replacing the joints with elastic components, which can be viewed as many very small joints. Repeating the cycle of application of the concepts of contradiction, resources, and improved ideality made big improvements in the ease of using the lamp. The evaluation criteria play a crucial role in the use of the improvement cycle—your table of "yes" and "no" answers will help you decide what aspect of the problem should be treated as the principal issue on the next cycle.

8.6 Summary

To evaluate any solution, first establish the evaluation criteria and then make the evaluation. The criteria should be, as much as possible, independent of the subjective feelings and interpretations of people. Different people should get approximately the same conclusions using the criteria. The basic concepts of TRIZ are translated into seven evaluation criteria that are used to evaluate proposed solutions with respect to the best conventional solutions.

When a problem is first solved, the evaluation may reveal drawbacks. Do not hurry to reject the idea. It may be poor, but it may also be excellent. It may only be necessary to solve subproblems in order to turn a marginal idea into a great idea. The first idea should nearly always be improved. There is a strong psychological barrier preventing the improvement of solutions. Repeating the cycle of application of the concepts of contradiction, resources, and improved ideality makes big improvements in the solution and helps overcome psychological inertia.

Chapter 9

Enriching the Model for Problem Solving

This book can be divided into three parts: (1) problem solving by analyzing contradictions, (2) the development of systems without direct use of contradiction analysis, and (3) the implementation of TRIZ for achieving business objectives. Chapters 2 through 8 covered the first part:

- Contradiction
- Resources
- Ideal Final Result
- Functional modeling

In this short chapter, we reach the midpoint.
Chapters 10 and 11 contain the second set of concepts and tools:

- Patterns of evolution
- Forty Principles of innovation

Chapters 12 through 17 address the implementation of TRIZ in organizations:

- How to identify problems to work on
- Using TRIZ for direction even when there are no visible issues
- How to implement TRIZ
- Use of TRIZ with various other tools
- Using TRIZ together with Six Sigma

Figure 9.1 shows a review of the three basic concepts studied so far. It is a more detailed presentation of Figure 8.1.

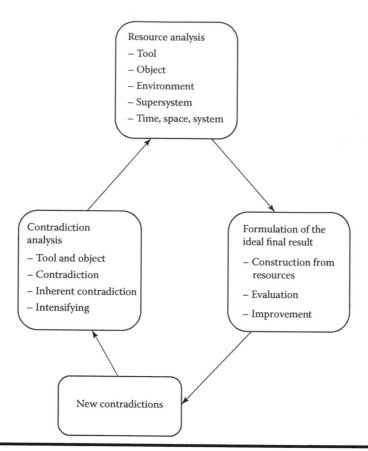

Figure 9.1   The cycle of TRIZ. The three upper boxes present three basic concepts. The lowest box shows the reformulation of contradictions. The loop shows that the problem-solving process is repeated. In practical situations, it is not unusual to go around the loop four or more times. You are making progress if you are answering different questions each time.

After the review, we present a general agenda for a problem-solving session. This is a one-page summary of the instructions given in previous chapters, which is illustrated with two familiar cases: lawnmower–noise muffling and carrot cultivation (Tables 9.1 and 9.2).

This agenda is a practical method that is easy to teach and to learn, and will get you started actually using the TRIZ methods fast. Many organizations that are looking for better methods for creativity have resisted TRIZ because of a perception that TRIZ is complicated and difficult. This impression has been due to the long, detailed, step-by-step guides for problem solving and problem-statement development in traditional TRIZ teaching systems. The most detailed step-by-step guide for problem solving is ARIZ. As explained in Chapter 2, ARIZ is a Romanized acronym of a Russian language phrase that, in English, stands for the algorithm for inventive problem solving. Altshuller and his team developed different versions of ARIZ between 1956 and 1985. A good review is published in Savransky's book.[1]

In this book, we present a short guide. We believe it is better to know a few things well, than many things superficially. A short problem-solving process can be easily repeated, and repetition or reiteration is important for mastering a new set of skills. As the tools and methods contained

Table 9.1   Summary of the Lawnmower Example

1.	Describe contradictions (Chapters 3 and 4)	
	a.	Describe the contradictions that make up the problem. There may be several on different system levels and in different stages of the life cycle of the product, process, or system.
		The lawnmower is too noisy. Contradictions: If noise is decreased, the lawnmower gets more complex. The lawnmower could be eliminated if we had grass that needs no cutting (such as moss), but to develop new grass may take much time and money.
	b.	Select one contradiction to resolve.
		If noise is decreased by making the muffler bigger, the lawnmower gets more complex.
	c.	Intensify the contradiction.
		Intensified inherent contradiction: The muffler should be present—the same muffler should be absent.
2.	Map resources (Chapter 5)	
	a.	List resources of the tool and object.
		Exhaust gas, noise. Try to use harmful elements as useful resources.
	b.	List resources of the environment.
		Air, gravity, the person who pushes the lawnmower.
	c.	List resources on the higher system level (macrolevel) and microlevel.
		Exhaust tube, lawnmower, grass, and soil.
3.	Define the Ideal Final Result (Chapters 6 and 8)	
	a.	Remove the contradiction using resources.
		Exhaust tube, exhaust gas, and grass do the job of the absent muffler. There is no muffler, but the noise vanishes.
		One technical solution: Grass as muffler.
	b.	Evaluate the solution.
		Conflict of big muffler–small muffler is solved.
	c.	Improve the solution.
		Imagine, for example, that the exhaust tube shape is changed (maybe the end segmented). The hot gas dries the grass more effectively and gently.

Table 9.2 Summary of the Carrot Cultivation Example

1.		Describe contradictions (Chapters 3 and 4)	
	a.	Describe the contradictions that make up the problem. There may be several on different system levels and in different stages of the life cycle of the product, process, or system.	
		The initial problem: Thinning carrots is an arduous job. This problem could disappear if we could plant seeds very accurately. But then we will have a new problem in another stage of the process: seeding precisely is difficult and time consuming.	
	b.	Select one contradiction to resolve.	
		The more precisely carrot seeds are planted, the slower the speed.	
	c.	Intensify the contradiction.	
		Intensified inherent contradiction: Very many seeds–one seed.	
2.		Map resources (Chapter 5)	
	a.	List resources of the tool and object.	
		Hand (guided by the eye and brain of the gardener), seed.	
	b.	List resources of the environment.	
		Soil, water, air, furrow, gravity.	
	c.	List resources on the higher system level (macrolevel) and microlevel.	
		Wastepaper, waste grass (from lawn).	
3.		Define the Ideal Final Result (Chapters 6 and 8)	
	a.	Remove the contradiction using resources.	
		Seeds, soil, and water make many seeds into one seed.	
		One technical solution: Biodegradable seed tape.	
	b.	Evaluate the solution.	
		Conflict between one seed–many seeds is resolved.	
	c.	Improve the solution.	
		Make the planting process even simpler by adding fertilizer to the seed tape.	

Table 9.3   Short Agenda for Problem Solving

1.		Describe contradictions (Chapters 3 and 4)
	a.	Describe the contradictions that make up the problem. There may be several on different system levels and in different stages of the life cycle of the product, process, or system.
	b.	Select one contradiction to resolve.
	c.	Intensify the contradiction.
2.		Map resources (Chapter 5)
	a.	List resources of the tool and object.
	b.	List resources of the environment.
	c.	List resources on the higher system level (macrolevel) and microlevel.
3.		Define the Ideal Final Result (Chapters 6 and 8)
	a.	Remove the contradiction using resources.
	b.	Evaluate the solution.
	c.	Improve the solution.

in this book are mastered, the student may then decide to move on to more advanced application topics.

Use the list in Table 9.3 as the agenda for a problem-solving meeting or as a guide for using TRIZ without meetings. We recommend using this kind of summary of your own examples. Table 9.3 is a template. In order to demonstrate how to use the template, we repeat the examples of the lawnmower and carrot cultivation (Tables 9.1 and 9.2).

Reference

1. Savransky, S. D. 2000. *Engineering of Creativity*. Boca Raton, FL: CRC Press, 304.

Chapter 10

Patterns Are Powerful Tools for System Development

10.1 Introduction

In Chapter 2, we introduced the patterns of evolution. The purpose of this chapter is to show how to use the patterns. In problem solving, knowing the patterns helps you to go from the features of the Ideal Final Result to concrete solutions and helps to visualize general solution directions that will move you toward the Ideal Final Result. Understanding the patterns helps you see how the system is evolving. If we see how the system is evolving, we can generate solution concepts; in this way, the solution can be developed without contradiction analysis, though the patterns can also help in solving contradictions. The place of patterns in the model for problem solving is shown in Figure 10.1.

Five of the most useful patterns of evolution are the following:

1. Uneven evolution of the parts and the features of the system
2. Transition to the macrolevel (supersystem) or incorporation in to the larger system of a higher level
3. Transition to the microlevel (subsystems) or the segmentation of the system into smaller parts
4. Increasing the interactions between systems.
5. Expansion and trimming of systems

These patterns are the five most common patterns though there are many others that can be further explored and utilized.[1,2] First, each of the five featured patterns will be explained, and then we will show how to use them both separately and together.

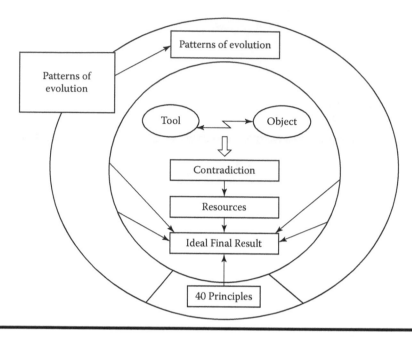

Figure 10.1   Patterns of evolution in the model for problem solving. Patterns are an independent tool kit helping to improve the system—arrows lead directly from patterns to the Ideal Final Result. They also support the problem-solving process from the contradiction to the description of the features of the ideal system.

10.2 Uneven Evolution of Systems

The uneven evolution of the system accounts for changes to different portions of the system at different rates over time. This uneven change to various subsystems allows for meeting some requirements of the operational environment, but can cause problems, bottlenecks, and contradictions elsewhere in the system. The unevenness concerns all systems and technologies: machinery, processes, organizations, and such. Unlike our typical way of thinking, particularly about technology, the evolution is not linear. There are always discontinuities, that is, incremental quantitative change chains are broken by qualitative leaps to new technology.

The history of the bicycle is a good example. In 1791 in France, de Sivrac developed the *celerifere,* which had two in-line wheels connected by a beam. The user "drove" it by pushing against the ground with his feet. This "hobby horse" technology was improved throughout the following decades, but there was a bottleneck in the system. The greater the speed, the more difficult it was to move the cycle by pushing off the ground. This contradiction was solved when the Michaux brothers added cranks and pedals to the front wheel's axle and created a bicycle boom with their velocipede.

Very soon a new contradiction appeared. The greater the speed, the more difficult it was to ride because the rider had to move his legs faster and faster. Increasing the diameter of the front wheel was the only way to get more speed from the direct-drive system.

The chain transmission, introduced in 1885, solved the problem of getting higher speed with smaller wheels. Then a new problem developed: the greater the speed, the more the vibration. Air-filled tires, which had been invented in 1845, resolved this problem. The bicycle reached the form it has today.

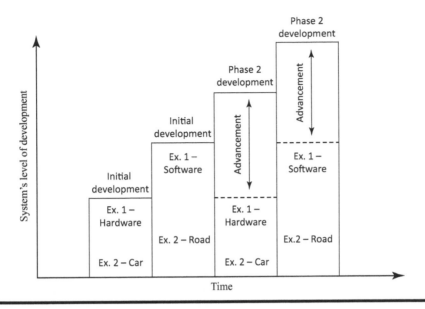

Figure 10.2   The uneven evolution of the systems.

Exercise: Think about a modern bicycle. Which contradictions can you name? How could they be resolved?

We see that usually some parts, or some features, improve rapidly, while other parts and features remain unchanged, sometimes for a very long time. Unevenness appears repeatedly and compels the system to evolve. The development of the car compelled the building of roads. Better roads made it necessary to develop better cars. (We could argue about whether it was necessary to improve the cars; or if it was then possible to sell better cars because there were roads for them, and if there is a market for something new, people will create a product for the market.) Analogously, computer hardware helps to make more effective software, and better software compels improvement in the hardware, as anyone knows who has purchased a new system only to find that it becomes obsolete within a seemingly short time. See the examples and exercise in Figures 10.2 and 10.3.

Exercise: Utilizing Figure 10.3, describe an example of uneven evolution of systems. It could be from your personal life or from your business life.

10.3 Transition to Macrolevel

The pattern of transition to the macrolevel describes a system that becomes better and better integrated into the higher-level system or macrosystem (supersystem). The system is not developing in a vacuum, as an isolated thing, but as part of, and merged into, a larger system.

The bicycle reached some important limits of development near the end of the nineteenth century. It was not possible to make significant increases in speed and transportation capacity of the human-powered vehicle. The bicycle was integrated together with the internal combustion engine into the higher-level systems. Motorcycles, cars, and airplanes developed. The motorcycle is directly a "motorized bicycle." The car also had its origin not only in horse-drawn carriages,

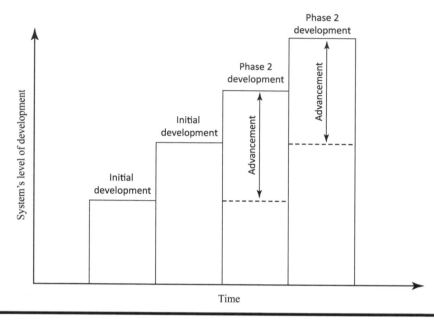

Figure 10.3   Exercise: Illustrate with your own examples the uneven evolution of the system.

but also in bicycles. The same goes for the airplane. After all, the Wright brothers were bicycle mechanics.

Exercise: What other ways can you suggest to integrate the bicycle into higher-level systems?

Stoves or fireplaces for heating a single room were developed to a high level a long time ago. To increase comfort and save time, stoves became central heating systems. In many northern countries, the integration has gone further. Large parts of cities are heated from a single thermal power station (so-called district heating, often with cogeneration of electric and thermal energy). Analogously, vacuum cleaners are combined into central vacuum cleaning systems. As discussed in Chapter 6, the kerosene lamp transitioned many of its subsystems into the macrosystem and only left the most important aspect of the lamp (the light-emitting device) in the newly transitioned system (the light bulb).

Clocks have been integrated into radios, television sets, cars, computers, mobile telephones, microwave ovens, and innumerable other systems. Other things are integrated, as well, such as electronic components, buildings, clothes (many layers), and many food products (e.g., the multilayer cake, the casserole with meat, vegetables, starch, and sauce).

Many examples of integration can be found in business, marketing, training, and other nontechnical fields. The entire financial field of mergers and acquisitions is a mechanism by which companies and other organizations are integrated frequently. Marketing is typically a system consisting of different media and ways of working (marketing mix).

The transition to the macrolevel is such a ubiquitous law that it is often dismissed as trivial. However, many problems that could be avoided arise because this law has been neglected. For example, in the 1980s, Apple's computers (Apple II, Lisa, and Macintosh) and Sony's Beta video systems met difficulties, although they were technically superior as isolated products. However, they were inferior due to the lack of integration into the larger system that the customer wanted to work with—the Apple systems did not integrate well with existing computing infrastructure and

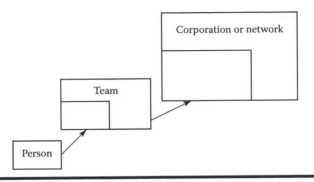

Figure 10.4   Transition to macrolevel.

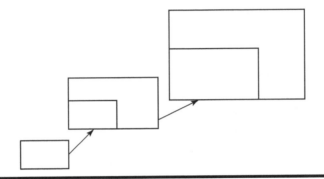

Figure 10.5   Exercise: Illustrate the transition to macrolevel with your own examples.

Beta tapes did not have enough recording time to completely capture the average made-for-TV movie.

Exercise: Describe an example of a transition to macrolevel. It could be from your personal life or from your business life.

See also examples and exercises in Figures 10.4 and 10.5.

10.4 Transition to Microlevel

The pattern of transition to the microlevel describes systems that are improved by dividing them into smaller and smaller parts. In Figure 10.6, three examples of the pattern are presented.

The first example has already been discussed briefly: replacing the solid edge of a cutting tool with a water jet. Molecules of water (and particles), instead of one solid object, do the cutting (see https://en.wikipedia.org/wiki/Water_jet_cutter).

Another example is from the field of health care. Early stretchers used to transport the injured were covered with simple canvas or some kind of mattress, which were not very good for carrying injured people with broken necks or backs because the person could not be held in a rigid position to prevent further injury. Vacuum mattresses were introduced to solve the problem. An airtight mattress filled with small plastic balls takes the shape of the body of the victim. After the injured person is laid on the mattress properly, air is suctioned from the mattress. A vacuum fixes the position of the balls with respect to the patient and each other, holding the patient securely during transportation. Many little balls replace the single solid support.

Examples:

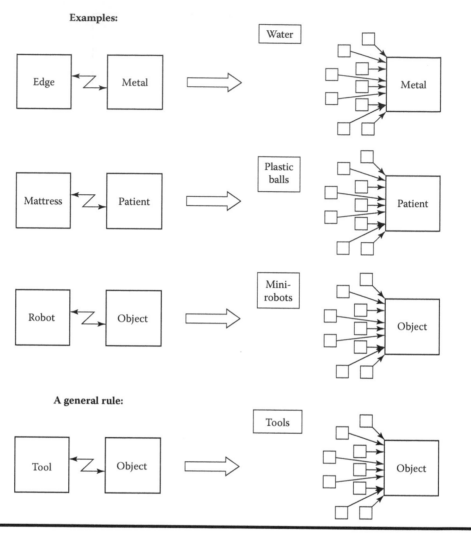

A general rule:

Figure 10.6   Three examples of the evolution pattern "transition to microlevel." Simple tool–action–object models can illustrate many patterns. On the left side is the system with a problem, on the right side the improved system.

A single big robot can only do a job such as digging or cleaning in one direction. It can be replaced by a quantity of mini-robots communicating with each other through radio or infrared waves. Many tiny robots replace the large one and can do the job from several directions simultaneously.

Some additional examples:

■ Car washes (washing machines for cars) frequently use brushes, but the brushes can scratch the car's surface. Now, water streams often replace brushes.
■ Cleaning cloths can go to the microlevel, too. So-called microfiber cloths may be so effective that washing chemicals are no longer needed.

- Stonewashing is a process used to produce the popular look of faded/worn denim. The method is improved by using enzymes instead of stones. This represents a transition to the microlevel in two ways:
  1. The enzyme molecules are much smaller than stones.
  2. The enzymes work on the fabric at the molecular level, whereas the stones act on the fabric at the level of the threads.
- The evolution of printing has gone through many generations of transition to the microlevel. Lithographic printing uses large (200 kg [440 lbs] or more) stones with all the letters etched on that single stone. Guttenberg's breakthrough was movable type, with each letter on separate pieces of small metal pillars. The matrix printer requires only a few (initially 9, later 24) tiny rods to make up each letter. The inkjet printer uses liquid ink and forms the letters from patterns of ink dots (initially 100 dots per inch, now as many as 600 dots per inch), and laser jets use light to sensitize the paper and fine powder to form the letters.
- Remember the firefighting example used in several chapters—a fine mist of water replaces liquid water.
- A classic example of the transition to microlevel is the manufacturing of glass on melted tin (Pilkington process). Big rolls of hot metal that were used to form the glass plate were replaced by a liquid tin bath used to float the glass plate.

There are three primary ways to segment material objects:

1. Segmentation of objects: Solid body, segmented body, powder, liquid, gas, plasma, and field. Most of the examples we have used so far are in this category.
2. Segmentation of space: Solid body, hollow body, multiple caverns, porous substance, and pores filled with an active substance. All kinds of spaces inside a body are frequently called "voids."
3. Segmentation of surface: Flat surface, corrugated surface, and rough surface.

Let us consider some more examples. We have repeatedly used the example of atomized water. If water can be segmented, why not air? In many processes, such as purification or flotation, air is mixed with water. Often, processes can be improved using smaller air bubbles or by using foam.

Transition to the microlevel can be used to some extent in business problems. Huge organizations are often segmented into many small independent organizations to get faster response to customer problems and faster product development. The "empowered employee" is a single person acting with the authority of the whole company—to fix problems and to take initiative much faster than the full infrastructure of the company would allow.

In the chapter on resources (Chapter 5), we used a serialized novel and television series as examples of segmentation in entertainment.

The lawnmower problem, first presented in Chapter 3, might also be solved by transitioning to the microlevel by utilizing smaller and smaller voids (highly porous material) as the noise-absorption medium used within the muffler.

The best solutions to problems often contain both the transition to macro- and microlevels. In microelectronics and communication technology, integrated circuits (microlevel) have enabled the building of global networks (macrolevel). The evolution of organizations has many analogous features. Small companies and subsidiaries build global networks.

Another interesting example of a simultaneous transition to the microsystem and the macrosystem might again be a solution concept for the lawnmower muffler problem. In the case of the

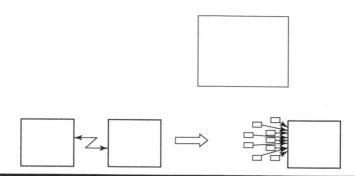

Figure 10.7   Exercise: Illustrate with your own examples the transition to the microlevel.

lawnmower muffler, it is possible to beam electromagnetic energy to the lawnmower and utilize an electric motor that does not require a muffler. The transition to the beamed energy represents a transition to the macrosystem (supersystem) by moving the energy source out of the lawnmower system and simultaneously represents a transition to the microsystem by moving from a chemical combustion energy source to that of an electromagnetic energy source.

Exercise: Describe an example of a transition to the microlevel. It could be from your personal life or from your business life.

See exercise in Figure 10.7.

10.5 Increasing the Interactions

Increasing interactions means adding new interactions or a transition to better-controlled interactions. This pattern also includes adding new substances that interact with the substances in the original system. "Substances" are materials, components, systems, and elements. They may be microorganisms (e.g., yeast), animals (e.g., bees), or humans (e.g., a hand as a tool). Further, they may be solid, liquid, gas, or field and can, therefore, take any form (carbon powder, screwdriver, oil, oxygen, electrons, or an electric motor to name a few). Substances interact with each other by a variety of means, including mechanical actions, thermal actions (heat, cold), acoustic interactions (different sounds), chemical reactions, electromagnetic fields and waves, odors, and biological interactions. You can find analogies in business such as with human interactions and communication.

The driver of the pattern of increasing interactions comes about as follows: the interaction between the tool and the object is functionally insufficient, excessive, or harmful (see Chapter 7). The system can be improved by adding new substances to the existing components, adding new interactions, or changing the substances and interactions in a variety of ways to amplify what is insufficient, reduce what is excessive, or eliminate what is harmful.

Consider some examples of the transition to better-controlled interactions (see Figure 10.8).

The interaction between the automobile and its environment is a big problem. New solutions are intensively developed that usually mean the introduction of better-controlled interactions. Navigation systems use radio waves, radar-equipped bumpers, drive-assistance systems with onboard video cameras, and so on. In 1997 in California, a magnetic control system was demonstrated. To improve automobile steering, magnetic pins were precisely located in the street. Sensors in the car detected the pins, and the steering was modified by the output of the sensor to keep the car from swerving off the road.

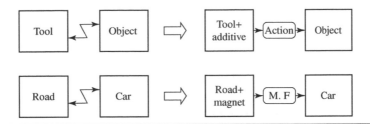

Figure 10.8   Increasing interactions. Mechanical connections are complemented or replaced by more controllable interactions. The mechanical contact between a car and a road can be complemented by electromagnetic interaction (M.F. [magnetic field]). Magnetic material is added to the road.

The history of the clock is a good example of the transition to more controllable interactions. The first clocks were sundials, which used the shadow of sunlight to indicate time. They could not be used on cloudy days or at night. Sand and water clocks and, later, pendulum clocks used gravitation. They worked day and night but were big and awkward. Spring clocks were introduced. They were smaller and easier to use but required winding. Modern clocks use vibrations of quartz crystals. The user cannot see the actual timekeeping mechanism. The history of the evolution of clocks demonstrates both the transition to the microlevel and the transition to more controllable interactions.

More examples:

- Post-it notes fixed by glue (adhesion) replace thumbtacks or pins (mechanical interaction).
- Barbed wire has a long history of improvement for enclosures for cattle. Now, low-energy electrified wire is used in many areas (electric interaction). A recent application of electric wire is to keep bears out of apiaries.
- Ultrasound, at frequencies that humans cannot hear, has replaced fences as a way to keep birds out of gardens. This is also an example of the pattern of segmentation, with a field replacing an object.
- Measuring devices have been transitioning to more controllable interaction states for some time now: wooden straight ruler, folding ruler, flexible cloth ruler, etc. Somewhat recently though, the measuring devise has also made the jump to the microlevel as represented by the laser ruler.
- The lawnmower contradiction might also be addressed by increasing the interaction of the engine-exhaust noise (sound waves) with the grass clippings. Might it be possible for the grass clippings to absorb most of the sound waves from the engine exhaust?

10.5.1 Simple Introduction of New Substances

The system can be improved by adding a new substance where a substance can be a raw material, a chemical, a component, or any other type of material object. To improve the performance of steel, carbon or nitrogen is added to the surface layer. To decrease the friction between the hull of an icebreaker and ice, a layer of polymer is added on the hull. Note that this pattern may violate the concept of the use of locally available resources because it may require the addition of substances that are not resources of the original system (see Section 10.6). The details of the specific situation will dictate whether new substances are needed or current resources can be used.

The wireless power system–improvement case study was introduced in Chapter 7. Several patterns of system evolution were applied to the original project; the example driven by the pattern of increasing interaction—introduction of new substances—is presented here.

For the wireless power system–improvement case study (see Section 7.7.1), the design was to place the transmitting antenna in the side panel of a laptop computer and the receiving antenna in the side panel of a mobile phone so that a user could charge her phone by simply laying it next to her laptop computer. The boundary condition of placing the sending and receiving antennas in the side panels of the laptop and mobile phone, respectively, resulted in limiting the effective diameters of the antennas and, therefore, the effective power transmission range (see Figure 10.9). Revisiting the initial problem statement, it will be recalled that the wireless power system did not operate well, or at all, if the receiving object (the mobile phone) was placed too far from the sending object (the laptop computer).

One pattern applied to this analysis was the pattern of increasing interactions. The pattern says that as systems become more ideal, the interactions between their components, or between their components and their environment, improve. The initial design of the wireless system included a single transmitting antenna in the side panel of a laptop computer and a single receiving antenna in the side panel of the smartphone (see Figure 10.9). One method of increasing the interactions between the transmitting (Tx) and the receiving (Rx) power antennas was to increase the number and locations of transmitting antennas. The simplest application of this concept was to utilize multiple identical antennas along the same side panel of the laptop. While this solution did not necessarily extend the distance between the laptop and the smartphone over which the power can be effectively transmitted, it certainly increased the possible locations alongside of the laptop over which the mobile phone can be placed and still receive whatever level of transmit power is available (see Figure 10.10).

10.5.2 Introduction of Modified Substances

Instead of a new substance, one can use a modification of the substances already existing in the system. To improve the performance of steel, the surface layer is quenched. To decrease the friction between the hull of an icebreaker and ice, water is added. The use of modified substances is closer to ideality than the use of new substances because modifications of the existing resources are used.

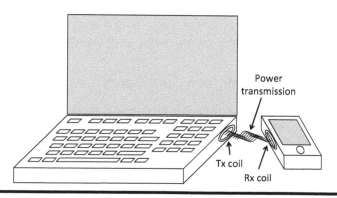

Figure 10.9   Wireless power antenna placement.

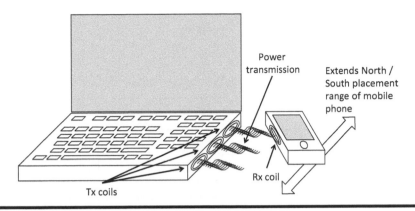

Figure 10.10   Wireless power with multiple Tx coils.

10.5.3 Introduction of a Void

Instead of a substance, one can use a void. It sounds odd to say, "instead of something, use nothing," but that is exactly what we mean. Examples include the following:

- Hollow structures instead of solid ones
- Foamed metal instead of solid metal objects
- Vacuum instead of antibacterial chemicals—a vacuum package
- Use of vacuum and suction in fixing, moving, and lifting

A "void" is anything more rarified than its environment.

10.5.4 Introduction of Action

Instead of substances and voids (things and nothing), one can use action. An example: dust can be removed in a cyclone-style vacuum cleaner using a mechanical action—centrifugal force. To improve performance, an electric field can be added to the cyclone.

In other literature on TRIZ, one can also find the term "field." The study of objects and interactions is called "substance-field analysis." In this book, we use the simple model tool–action–object, or tool–interaction–object. The term "interaction" covers both fields in common language (electromagnetic fields, gravitation) and interactions (chemical, thermal, mechanical, and biological) that are usually not called fields. One can speak also of social and human interactions. The concept of interaction is accurate and helps the TRIZ user to see opportunities to use different interactions.

Exercise: Describe an example of the increase of interactions. It could be from your personal life or from your business life.

Also see the exercise in Figure 10.11.

10.6 Expansion and Trimming

The last pattern we will consider is called "expansion and trimming," or sometimes "pulsating" evolution. The system expands first, becoming more complicated, then it is trimmed (or convolutes); that is, its elements are combined into a simpler system. The increasing number of parts

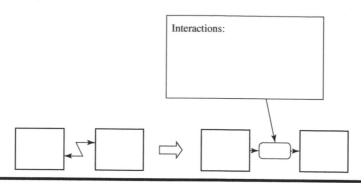

Figure 10.11 Exercise: Illustrate increasing interactions by your own example.

and operations cause problems that are solved when the system is simplified. The evolution is not linear. We can say that the system "pulsates." First, there are few parts and operations. Then, the number of components and operation stages grow quickly until the system "collapses" and is trimmed to a few parts. Then, the cycle begins again.

Transistors and other microelectronic components were first used as single parts. Systems with many components became very complex, with short lifetimes. Later, a great number of small components were combined into the integrated circuit, which can be treated as a single component. The electronics industry has been through many cycles, combining integrated circuits into complex systems, and then simplifying the system by making components with higher levels of integration. Integrated circuits became a new monosystem that is further embedded into other systems.

However, the pulsating pattern does not only occur in high-tech electronics. Traditional bicycle wheels have a great number of spokes, but new wheels using disks, or other means of attaching the wheel to the hub, have been introduced recently.

The microfiber cloth, used as an example of the transition to the microlevel, is also an example of expansion and trimming. The system of cloth and detergent is trimmed to a microfiber cloth.

The car tire was first improved by adding an inner tube. Later, the inner tube was removed. The spare tire was a staple in automobiles for many years. Now, it is removed in some cars and the tire itself works as the spare. In heavy trucks, double tires were introduced. There are now concept designs for a single, very wide tire.

Expanding and trimming also improve processes. Water-jet cutting often makes it possible to merge cutting and machining because the surface does not need any machining after water cutting.

Printing technology is evolving from five processes to two: standard methods require preparing the content, making the film, making the plate or cylinder, fixing the plate/cylinder to the printing machine, and printing. The new technology consists of two phases: preparation of the content and printing.

See, as well, the citation from Toynbee in Chapter 6: transportation technology expanded first—locomotive—and then was trimmed—internal combustion engine. Long-distance communication technology first expanded—telegraph and telephone wires—and then was trimmed—wireless technology. Clothing expanded from primitive furs and fibers to complex dress in the seventeenth century and then was simplified (tee shirts and casual pants).

This pattern can also be called "mono-bi-poly," because a monosystem is combined with another to form a bisystem, and then more are added to form a polysystem.

Other examples of polysystems are

■ Coffee maker with a grinder, clock, and steamer
■ Multiple colored pencils in a set
■ Multiple bank tellers
■ Smartphone (multifunctionality)

When the polysystem is simplified, it can become a new monosystem. Salamatov has analyzed the pattern mono-bi-poly in detail.[3]

Both similar and dissimilar systems can be combined. Examples of similar and dissimilar polysystems are

■ Multiple colored pencils in a set (similar)
■ Coffee maker with a bean grinder, clock, and steamer (dissimilar)
■ Multiple bank tellers (similar)
■ Toolbox: Hammer, screwdriver, tape measure, level (dissimilar)
■ Safety glass: Glass layered with transparent plastic (similar)
■ Smartphone: Phone, camera, Internet device, etc. (dissimilar)

We have already introduced many examples of the mono-bi-poly transition. Fragile glass plates can be handled more easily if they are packaged together. Juice packages can be moved more easily if they are affixed to each other. The word "sandwich" meant at first only slices of bread with a filling. Soon people began calling anything that had multiple layers a sandwich. Sandwich structures are used in buildings and airplanes. Textiles and cooking often have many layers. Glass is often sandwiched with other materials, which in fact is more common than using multiple glass layers. As mentioned, one way to make safety glass is to sandwich plastic between layers of glass.

In the case of layered systems (similar or dissimilar), the pattern of mono-bi-poly could be called the sandwich pattern. The combination of similar systems is a special case of sandwich principle.

Exercise: Describe an example of expansion and trimming. It could be from your personal life or from your business life.

See the examples in Figures 10.12 and the exercise in Figure 10.13.

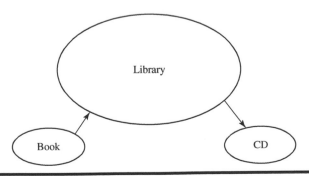

Figure 10.12 Expansion and trimming.

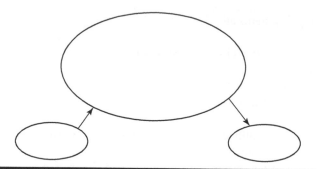

Figure 10.13   Exercise: Illustrate expansion and trimming by your own examples.

10.7 How to Use Patterns Together

Patterns should be studied together and the result checked by the criteria of the Ideal Final Result. Considering only one single pattern can often lead to incorrect ideas about possible patterns of evolution. Thinking about multiple patterns together provides more reliable results. The authors

Table 10.1   A Summary of Patterns

Pattern	*About the Pattern*
Uneven evolution of the system.	Uneven evolution of parts. Uneven evolution of process stages. Uneven improvement of features. Repeating rise of unevenness.
Transition to macrolevel.	Combining one system with a similar or a dissimilar system, or with many similar or dissimilar systems (mono-bi-poly). Transition to macrolevel is repeated.
Transition to microlevel.	Solid body, segmented body, liquid or powder, gas or plasma, field. Solid body, hollow body, many caverns, porous substance. Flat surface, corrugated surface, rough surface.
The increase of interactions: introducing substances and actions.	Introducing substances: new and modified substances, void. Introducing actions: mechanical, acoustic, thermal, chemical, electric, magnetic.
Expansion and trimming.	Increasing number of parts. Increasing number of operations. Trimming to fewer parts and operations. Cycles of expansion and trimming are repeated.
Increasing of the ideality of the system.	One pattern is used to increase ideality. If the use of one pattern causes new problems, other patterns are used to resolve them. Many patterns are used.

have found that each pattern used produces solution concepts that are comparable to pieces of a puzzle. When the "pieces" of ideas are combined, they produce a clearer and more compelling solution vector than when the patterns are applied independently. Drilling down even more, each pattern can produce completely different ideas or can produce ideas with similar themes. Both outcomes are useful as different ideas can be combined into a grand concept, and similar ideas can be used to highlight solution directions that, due to their repeated appearance, are likely best choices. Also, it has been found that independent ideas from different patterns may be vague or nebulous when considered alone but obtain clarity when looked at through the combined lens of all the patterns' output. In other words, if a pattern generates an idea, go ahead and capture it even if at first it is not clear how it would be applied or exactly what it implies about the solution direction.

For the practical application of the patterns, we suggest using "five + one patterns": the five patterns we have just discussed plus the pattern of increasing ideality. There are additional sets of evolutionary patterns documented throughout the TRIZ literature (two such collections can be found in the book *TRIZ for Dummies* by L. Haines-Gadd[4] and "Systematic value innovation approach" by T. Kanikdale[5]), but as an introduction we suggest using this set of "five + one." Table 10.1 summarizes five + one patterns and some information about the contents of the patterns.

10.8 Benefits from Understanding the Patterns of Evolution

We can name three uses and benefits of the patterns:

1. Use the patterns of evolution as tools for the evaluation and selection of solutions to problems. They complement the ideality criteria evaluation. Some good questions to use in reviewing a proposed solution to a problem are the following:
   Does this solution demonstrate the uneven evolution of the technology? What part will need improvement next?
   Will the next system transition be to the macro- or microlevel? Or, will parts of the system transition to the macrolevel while others transition to the microlevel?
   How will interactions increase?
   How does the ideality of the system as a whole increase?
2. Use the patterns of evolution as aids in the identification of opportunities. More specifically, use the patterns to characterize a current manifestation of a system and then the same patterns to project next-generation advancements. Examining each pattern can give you information. What are the spearheads and bottlenecks in the system's evolution? How can you integrate the system into the next higher-level system? How do you segment it into smaller parts? How can you increase interactions? One can create what-if studies of the future of the systems evolution. What can be achieved if the innate potential of the technology is used? When will it be necessary to use a different technology to improve the ideality of the system?
3. The use of the solutions from other industries gets easier. It will be easier to see similar features and use them. For example, the segmentation and integration principles that are used in electronics and equipment manufacturing can be transferred to the building industry. Patterns of self-service in retail business are repeated in education and in medicine.

10.9 Some Nuances in the Use of Patterns

We have presented the patterns in simple ways and omitted many details. However, to avoid over-simplification, keep these points in mind:

- Uneven evolution of the system, the transition to macrolevel, and pulsating evolution are the most universally recognizable patterns. By "universal" we mean that, in almost every case, we can easily see these patterns and use them to develop the system.
- The pattern of increasing interactions is perhaps the most consistent of those considered. For example, systems contain more and more electric and magnetic interactions. The trend has been steady for more than 100 years.

Technologies develop incrementally over time and can both stand alone and augment other previously developed technologies. These "Floors of Technology" can be used as a rule of thumb when forecasting changes to systems. It helps to remember that technologies are very often built on each other, as are the floors of a building. Ancient mechanics continue to exist as a first floor (see Table 10.2). Very often one floor (e.g., mechanics) continues to live in concert with a newer floor (e.g., electromagnetic fields). As an example, optomechanics is one of the emerging technologies of today. The "Floors of Technology" rule does not mean that every technical system should contain all floors. It says that if you have a system, you might often develop it into a better system by adding some new "floors."

There are also multiple sources of energy that can be used within a solution concept, and these energies are closely related to the "Floors of Technology" discussion. "MATChEM" is a mnemonic device to help trigger the consideration of fields for use in problem solving. The mnemonic stands for mechanical, acoustical, thermal, chemical, and electromagnetic fields and

Table 10.2 Floors of Technology

Baseline of the Floors of Technology	Examples of Typical Innovations
Mechanics	Lever, wheel and axle, pulley, inclined plane, wedge, screw
Acoustics	Environmental noise and soundscapes, musical acoustics, ultrasonics, vibration, and dynamics
Chemistry	Chemical formulation, chemical battery, gas relative pressure, H_2O, chemical symbols, atomic weight, chemical bonds
Thermal science and technology	Steam engine
Electric technology and industry	Light bulb, coal chemistry, machine building with replicable parts
Microelectronics and industrial acceleration	Radio technology, microprocessor, mass production, fission energy, microbiology
Photonics	Nanotechology, solar energy, advanced laser, nano-bio-info-cognition, biophotonics
The "next" floor (maybe Gravitonics?)	(Levitation, gravity tractor beams, etc.?)

interactions (MATChEM is also discussed in Sections 5.5.7 and 7.7.3). However, this does not mean that mechanical interactions should always be replaced, or compete, with electromagnetic fields. The pattern says that the transition to more controllable fields happens so often that the utilization of a change of interaction energy category should be considered. The transition to the microlevel also happens frequently. Some few exceptions are found. For example, sometimes chemical washing is replaced by mechanical cleaning, or cleaning by water or steam to get rid of chemicals. One should remember the probabilistic character of patterns and check whether the changes implied by them increase the ideality of the system. Each of the fields represented by MATChEM can be used by themselves or in conjunction with other fields. Further, each field interacts with matter on a different level (see Table 10.3). This interaction level provides flexibility in application based on the system's needs and relates directly to the pattern discussed in Section 10.5.

One frequently asked question is how there can be any accurate patterns or laws in the evolution of systems, when most predictions of the future, and society, are very unreliable. If the patterns are true scientific laws, should we not be capable of precisely predicting system evolution, or at least the evolution of technology? We can predict the technology of the future with the empirical accuracy that is significant enough for practical purposes. We can predict, for example, that humans will land on Mars before 2050. It is not as precise as the astronomical calculation of the forthcoming eclipse of the moon, but it still gives useful information. One prediction, shown in Section 10.10, was that information and communication connectivity was going to grow exponentially. While that forecast does not give specifics as to how that prediction will manifest itself in the marketplace, it does provide valuable insight into the development directions for computing, communications, and other connected appliances. There is also the time factor. In other words, the farther you go into the future, the more accurate the application of patterns of evolution becomes. The authors' experience is that when applied properly, the patterns create very accurate forecasts as far as what will occur. What is not determined by the application of the patterns of evolution is when will the change transpire and who, or what, will drive that appearance.

In the beginning of the book, we presented examples of "late" innovations. Obviously, innovations such as penicillin, fast-food restaurants, and flash melting of metals could have appeared some time earlier or later than they really happened. In the long run, the evolution was inevitable. One can accelerate or retard the change, but not prevent it.

Table 10.3 MATChEM and the Interactions

MATChEM	*Interaction Abstraction*
Mechanical	Object
Acoustic	Surface
Thermal	Lattice/Matter
Chemical	Molecular
Electromagnetic	Electron/Spin

10.10 Examples of the Application of Patterns of Evolution

The application of the patterns can be used to forecast new technological advancements. As a quick example, over the last more than 10 years, one of the authors wrote several articles and stories that included the prediction that videophones would become commonplace in homes. The forecast was made clear to the author based on the application of the patterns of evolution. Those stories were published between the years of 2003 and 2006. Then, somewhere around 2010, the web-based tool Skype became so commonplace that the "video phone in every home" forecast had essentially come true.

New solutions can be obtained and systems improved by applying the patterns of evolution.

Tables 10.4 and 10.5 summarize the application of the patterns of evolution to the lawnmower noise-reduction problem and the carrot cultivation examples, respectively. Table 10.6 is a template for the practice of the technique.

This section also contains abbreviated presentations of three different case studies of application of the patterns that the author performed for clients between the years of 2010 and 2015. A careful study of these three case studies (two products and one service) will provide deeper insight as to the application of the patterns of evolution. After you have examined these studies in detail, utilize Table 10.6 and practice the technique for yourself.

10.10.1 Case Study One: How Will Smartphones Evolve?

In 2010, the author performed an analysis of the future state of smartphone technology and capabilities for a leading smartphone manufacturer. As discussed in Sections 10.7 and 10.9, the general technique is to understand the current state of the product in relation to the various patterns and then forecast what further movement along the patterns may look like. Reviewing Table 10.7, the thought process that went into each advancement, as proposed by the patterns, is shared, followed by an update as to how well that evolutionary forecast held up. Keep in mind that this case study was executed in 2010 (7 years prior to the publication of this book).

Table 10.4   Decreasing the Noise in the Lawnmower

Pattern	How to Apply to the System
Uneven evolution of the system.	Noise suppression, decreasing pollution.
Transition to macrolevel.	Combination with other machines. Solving the problem at the level of the garden system.
Transition to microlevel.	Porous materials.
The increase of interactions.	Noise against noise. Sensing the sound waves and generating waves that cancel them.
Expansion and trimming.	Increasing and decreasing number of parts. First, making the muffler bigger and, later, eliminating the muffler altogether.
Summary: Increasing of ideality.	Cleaner lawnmower. Less lawnmower. Absent lawnmower. Grass stays short by itself.

Table 10.5 How to Cultivate Carrots

Pattern	How to Apply to the System
Uneven evolution of the system.	Developing advanced nutrients for the carrots while keeping all other traditional farming techniques the same. Using the experience of growing vegetables in closed rooms.
Transition to macrolevel.	Adding fertilizers to the tape. Cultivating carrots in a closed, clean container.
Transition to microlevel.	Cloning carrots in lab, as opposed to growing in garden.
The increase of interactions.	Using reflective surfaces to increase incident sunlight. Utilized in ground "pots" to concentrate water where needed.
Expansion and trimming.	Build a fully automated electronically activated watering system and then trimming to simple rainwater gravity feed system with valves actuated by ambient temperatures.
Summary: Increasing of ideality.	Improved planting, watering, feeding and growth.

Table 10.6 Select One of Your Problems and Apply the Patterns to It

Pattern	How to Apply to the System
Uneven evolution of the system.	
Transition to macrolevel.	
Transition to microlevel.	
The increase of interactions.	
Expansion and trimming.	
Summary: Increasing of ideality.	

Uneven evolution of the system. There are a wide variety of ways to consider and apply this pattern, but the author often focuses on portions of the system that seem to be lagging. For example, computational power, photo quality, data-transfer bandwidth, and other features of smartphones seemed to have been moving along just fine. What wasn't keeping up was the ability for phones to have large displays without being accompanied by a large form factor (overall phone size and shape). The trend of making phones larger and larger to have larger displays does not address the contradictory requirements between display size and phone size. It is desired to have a large display, but it is not desired to have a large phone. Therefore, alternative display methods were forecast: projection, eyeglass displays, and retinal imaging (projecting images directly into the user's eye to create a "display" as large as needed without the need for an actual screen).

Current state of smartphones in relation to prediction:

■ Google Glass introduced to the market in 2013
■ Author supported large Korean electronics manufacturer in developing 3D projection imaging (2014)

Table 10.7   Patterns Applied to Smartphone (2010 Case Study)

Pattern	General Concept	Application
Uneven evolution of the system.	Focus on increasing size of display (without increasing the form factor size).	Utilize image projection or retinal projection.
Transition to macrolevel.	Further connect smartphone to its operational environment.	Allow smartphone to control and receive data from new systems.
Transition to microlevel. • Communication field	• Move from communication field to noncommunication field	• Incorporate active jamming into smartphone capabilities
Increase of interaction. • Introduction of an action	• Increase level of connectivity	• Allow phone to connect with any and all operational environment components
Expansion and trimming.	Add more operational capabilities.	Have smartphone completely control itself.

■ Retinal projection featured in *MIT Technology Review* (March 2015 issue)

Transition to the macrolevel and increasing of interactions/introduction of actions: How can the system be better integrated and combined with the greater operational environment? At the time of the analysis, smartphones were mostly connected to each other (phone service) and websites (Internet providers). The forecast was to connect the phones to any system that would provide value to the user such as home (heating/cooling, lights, security systems, etc.), bank, pets, automobile, traffic lights, soda machines, and cash registers.

Current state of smartphones in relation to prediction:

■ Control your home with your smartphone (thermostats, security cameras, security systems, pet feeders (2014), lighting, appliances, smart keys (2013), electrical outlets, etc.): 2011 and on
■ Introduction of item tracking systems (e.g., Tile): 2013
■ Phone-based payment systems (e.g., Apple Pay): 2014 (piloted 2011)

Transition to the microlevel: One aspect of this pattern is the type of field involved in the microlevel interaction. One of the last stops along this trend is the incorporation of interference fields (anti-fields). This led the author to consider active blocking (pseudo jamming systems). In other words, increase the interactions between the smartphone and hostile systems by actively blocking them.

Current state of smartphones in relation to prediction:

■ Blackphone (high-security smartphone): 2014

Expansion and trimming: One aspect of the expansion portion of this pattern was to place more of the system requirements (transmission, energy source, control and decision making) into the smartphone. While the smartphone was already fairly far down this path, adding more and more decision-making capabilities to the smartphone would continue this journey. Over the past several years, smartphones have become more and more self-sufficient.

Current state of smartphones in relation to prediction:

■ Apple's Siri (introduced in 2010).
■ Many copycats followed (2011 and on).
■ New versions of Google Pixel phone automatically send you calendar events from organizations you are associated with, remind you to check in for flights, etc. (2016).

10.10.2 Case Study Two: How Will Delivery Services Evolve?

In 2013, one of the author's clients (global freight service) requested an analysis as to how delivery services would evolve. Reviewing Table 10.8, the thought process that went into each advancement, as proposed by the patterns, is shared, followed by an update as to how well the evolutionary forecast held up. Keep in mind that this case study was executed in 2013 (4 years prior to the publication of this book).

Uneven evolution of the system and transition to the macrolevel and microlevel: There are a wide variety of ways to consider and apply these patterns, but the author often focuses on the primary function (main function) of the system under analysis. In the delivery service case study, the main function of the service is to *move* things. Therefore, what is a more effective and efficient way to move things? The Internet is very effective at moving information in a very economical manner. Further, data about an item is one of the most representative aspects of an item and can easily be moved by the Internet. Using the Internet represents both a macrolevel (supersystem merger) and microlevel (field transmission) transition.

Table 10.8 Patterns Applied to Delivery Services (2013 Case Study)

Pattern	General Concept	Application
Uneven evolution of the system.	Focus first on main principle of action of the system.	Change the way in which the items are moved.
Transition to macrolevel.	Incorporate movement of goods into other systems.	Use Internet to move information about the goods.
Transition to microlevel. • Delivery "vehicle"	• Move from rigid jointed system to field system	• Deliver goods by way of digital information
Increase of interaction. • Introduction of new substance • Introduction of action	• Raw materials • Build new item at point of delivery	• Provide raw materials at point of need • Utilize manufacturing at point of need
Expansion and trimming.	Add in manufacturing equipment at point of need and remove as much of the existing components as possible.	Utilize 3D printing.

Increase of interaction:

- Introduction of new substance: If only information about an item is transferred, then something must be done with that information once it arrives at its destination. Since the goal is to have the item "appear" at a new location then it makes sense that, at the minimum, it would need to be materialized (or rather manufactured) at its point of need. In this case, it would be necessary to have raw materials available at the "shipping" destination.
- Introduction of an action: In direct relation to the discussion on introduction of a new substance, it will be necessary to perform an action with that new substance (raw materials). For example, develop manufacturing capabilities at the "shipping" destination.

Expansion and trimming: In concert with the concept of manufacturing at the point of need by utilizing digital information about the item to be "shipped," the use of 3D printing seemed like a natural fit and represented the expansion portion of this evolutionary pattern. Further, once 3D printing is available then the original shipping equipment, trucks, airplanes, etc. can be trimmed.

In summary, the delivery services prediction states that as 3D printing capabilities increase then more and more items can simply be printed at their point of need. This reduces the need for shipping specific items and only requires that the digital information and raw materials be shipped. Raw materials can be packed and moved more efficiently than discreet items (can even be pumped through a pipe in many cases). Residential and commercial printers are already available and their capabilities will increase exponentially over time (UV-triggered photopolymerization appeared in 2013). Fundamentally, 3D printers are the entry-level engineering systems that are driving the manifestation of true teleportation systems envisioned by way of "replicators" in science fiction shows such as *Star Trek*. This says that the commercial shippers will need to focus on three topics:

1. Enter the digital information ownership business model.
2. Invest in large regional 3D printers to take care of items that cannot be printed by smaller residential or commercial system (e.g., automobiles or farm tractors) and then provide local delivery services of those larger items.
3. Provide long-distance delivery service for the raw materials necessary for use by the 3D printers (metals, plastics, organic compounds, etc.).

Current state of delivery services in relation to prediction:

- Major players in printing are developing large-scale commercial printers (2015).
- 3D-printing capabilities are becoming more and more capable (within 1 year (2014) of the initial study, the following 3D print capabilities came into existence):
  - 3D printing of an operational human kidney (USA)
  - 3D printing of a working jet engine (Australia)
  - 3D printing of advance and lightweight soccer cleat (Nike)
  - 3D printing of a two-story house (China)

10.10.3 Case Study Three: How Will Clothes Washing Evolve?

In 2015, one of the author's clients (an Asian manufacturer of electronics and appliances) had seen TRIZ predictions by a European competitor (Whirlpool, Italy), presented at the European TRIZ

Association (ETRIA) international TRIZ conferences, and wanted the benefit of a TRIZ perspective. The client was concerned with how clothes washing systems (i.e., washing machines) would evolve. Reviewing Table 10.9, the thought process that went into each advancement, as proposed by the patterns, is shared, followed by an update as to how well the evolutionary forecast held up. Keep in mind that this case study was executed in 2015 (2 years prior to the publication of this book).

Uneven evolution of the system: There are a wide variety of ways to consider and apply these patterns, but the author often focuses on the primary function (main function) of the system under analysis. In the clothes washing case study, the main function of the engineering system is to *remove* dirt. Therefore, what is a more effective and efficient way to *remove* dirt? The desire is to improve the removal of dirt.

Transition to the macrolevel: While we are at it, why not incorporate the new systems into other operational environment components? Where are clothes stored while not being worn? In a closet or drawer. Try to design the new system to operate where the clothes are; thus, eliminate the need to move the clothes to, and from, the washer.

Transition to the microlevel:

- Drum and agitator: Since the drum and agitator are currently a rigid/jointed system, moving them to the state of a flexible or even gaseous system would be advisable. The drum and agitator should become gaseous.
- Washing media: The current washing media (water) is a liquid. The next steps along this path are said to be gases or fields. The washing media should become a gas.

Table 10.9   Patterns Applied to Clothes Washing (2015 Case Study)

Pattern	General Concept	Application
Uneven evolution of the system.	Focus first on main principle of action of the system.	Change the way in which the soil is loosened.
Transition to macrolevel.	Incorporate washing machine into other parts of the home.	Place new system in closet of bedroom.
Transition to microlevel. • Drum/agitator • Washing media • Detergent (surfactant)	• Move from rigid jointed system to flexible or gaseous system • Move from water to gas or field • Move from liquid soap to gas or field	• Use gas as washing drum • Make washing media a gas • Use ionizing radiation as a surfactant
Increase of interaction. • Introduction of new substance • Introduction of action	• New surfactant • New removal of soil	• Ionizing radiation (surfactant) • Gas to carry away loosened soil
Expansion and trimming.	Remove as much of the existing components as possible.	Gas to carry away soil. Eliminate drum, agitator, motors, door, body, etc.

■ Detergent (surfactant): The current system detergent (surfactant) is a liquid in most cases. The next steps along this path are said to be gases or fields. The detergent should become a field such as an ionizing field to interact with, and loosen, the soil. By the way, several years before the publication of this book, there were already commercial applications of using gaseous and ionizing field-radiation surfactants.

Increase of interaction:

■ Introduction of new substance: Add an ionizing field.
■ Introduction of an action: Using an ionizing field to interact with the soil represents a new action (with better and more microlevel interactions), and the use of a gas to carry away the loosened soil also represents a new action (likewise with better and more microlevel interactions).

Expansion and trimming: Adding a gas to remove the soil allows the elimination of the drum, agitator, and motors, while the application of the new system where the clothes already are (the closet) allows the removal of the current systems structure such as the body and door.

In summary, the new system will be built into a clothing closet and utilize ionizing radiation to loosen the soil from the clothing as it hangs in the closet (no need to move clothing to, or from, the "washing machine"). Next, the loosened soil will be removed from the clothing by way of a gaseous cloud.

Current state of delivery services in relation to prediction: This may seem like science fiction (even in 2017). However, there are already several advancements/systems that make this forecast much closer to reality than we might think:

■ NASA: Cold water, no-detergent laundry technology: 2008[6]
■ Water-based, ionizing, and air-drying clothes-washing system (India): initially developed 2008[7]
■ Lasers to eliminate stains: 2010[8]

10.11 Summary

The same patterns are repeated in the evolution of systems. These patterns can be used for the further development of the system.

There are five primary patterns of evolution: uneven evolution of the technology, transition to macrolevel, transition to microlevel or segmentation, increase of interaction, and expansion/trimming (convolution).

You can use the patterns of evolution for selecting solutions, finding or solving problems, forecasting evolution, and transferring solutions across industries.

References

1. Zlotin, B. and Zusman, A. 1999. *Tools of Classical TRIZ*. Farmington Hills, MI. Ideation International. http://www.ideationtriz.com/publication.asp#Tools_of_Classical_TRIZ ISBN 1-928747-02-7.
2. Zlotin, B. and Zusman, A. 2002. *Directed Evolution: Philosophy, Theory, and Practice*. Farmington Hills, MI. Ideation International. http://www.ideationtriz.com/publication.asp#Directed_Evolution ISBN 1-928747-06-X.

3. Salamatov, Y. 1999. *TRIZ: The Right Solution at the Right Time*. Hattem, the Netherlands: Insytec, 192.
4. Haines-Gadd, L. 2016. *TRIZ for Dummies*. Chichester: John Wiley & Sons, 61–81.
5. Kanikdale, T. 2016. Systematic value innovation approach. *The TRIZ Journal*, May 2. https://triz-journal.com/systematic-value-innovation-approach/.
6. Albrecht, S. 2008. New, NASA-developed, cold water. *Healthy Beginnings*, July 4. http://hbmag.com/new-nasa-developed-cold-water/.
7. Vardhan, H. 2017. An alternative clothes cleaner cum lounge seat. *Elite Choice*. http://elitechoice.org/2008/07/09/an-alternative-clothes-cleaner-cum-lounge-seat-by-harsha-vardhan/.
8. Hines, M. 2010. The Supernova Robot uses lasers to eliminate stains. *Trend Hunter Eco*, September 2. http://www.trendhunter.com/trends/supernova-robot.

Chapter 11

Principles for Innovation: 40 Ways to Create Good Solutions

11.1 Introduction

We have repeated throughout this book that understanding the common attributes of good solutions is crucial to the enhancement of creativity. One obvious conclusion is to simply make a list of the most important attributes and then use the list for generating bright ideas and successful products.

Good generic solutions across industries have been studied as part of the development of TRIZ. Altshuller and his researchers collected examples of repeated use of the same solutions from patent information. After painstaking work, the information on tens of thousands of good solutions was boiled down to 40 Principles in the early 1970s.[1] The use of these principles of innovation became an important branch of the theory. Various collections of standard solutions and principles for innovations were developed. To learn more about the research, see Savransky's book[2] and the paper by Zlotin and Zusman.[3]

Our goal in this chapter is to present the modern list of 40 Principles as a problem-solving tool that is effective, easy to use, cheap, and accessible to everybody. The 40 Principles are the most popular tool of TRIZ and several books and many articles are devoted to them. This version has three main aspects:

1. All examples are new. They are mainly examples of innovations or realized solutions that are actually used in everyday life.
2. Most principles are illustrated with examples from both business and technology.
3. The principles are presented in a compact form, without division in to sub-principles, as in early books. One can manage 40 Principles much easier than 80 to 90 sub-principles.

Some of the aspects of earlier versions of the 40 Principles are preserved in this version:

■ The structure of the list of the principles, that is, the number (40—not less, not more) and the order of principles are the same as in older books, so this book is compatible with older publications.

■ The names or labels of the principles are also conventional. For example, one principle is named "strong oxidants." We show how, for some situations, this can mean the use of strong emotions, as well as the use of particular chemicals, but we keep the classical name for the principle.

Research continues and there may possibly be more than 40 Principles at some time in the future (or fewer, if the list is reorganized.) For now, the 40 Principles are the standard and new examples and new ways of using them have been added to expand their use. If you would like to get additional examples to help you solve your problems, the 40 Principles, with examples from various disciplines, can be downloaded from http://innomationcorp.com/simplifiedTRIZ. These selections are authored by a variety of individuals (including the author) and, as a result, show a range of styles and quality (i.e., some of the examples you may like, other examples might feel forced or poorly developed). Regardless, you will surely get new viewpoints regarding the application of the 40 Principles. Among the 40 Principles listed for various applications are the following:

■ Architecture
■ Business
■ Chemical engineering
■ Chemistry
■ Construction
■ Computing
■ Customer Service
■ Eco-innovate examples
■ Education
■ Food industry
■ Marketing, sales, and advertising
■ Microelectronics
■ Quality management

Go to http://innomationcorp.com/simplifiedTRIZ to download these and other documents.

The structure of this chapter is simple. First, we will introduce the principles with some examples. Then we will show how to select principles that will help solve particular problems, including the Contradiction Matrix as an important tool in support of the search.

Figure 11.1 shows the place of the principles in the model for problem solving. You can use the 40 Principles as an independent tool. This chapter is written so that you can read it and get benefits from it without reading the other chapters of the book.

The efficiency of the 40 Principles increases when they are used together with other tools. As an example, function analysis helps direct the application of the 40 Principles to the area of conflict within the system under analysis. Patterns of evolution, ideality, and contradiction analysis may give the same solutions or may give different solutions from the 40 Principles—considerable overlap is quite common. The tools strengthen and enrich each other.

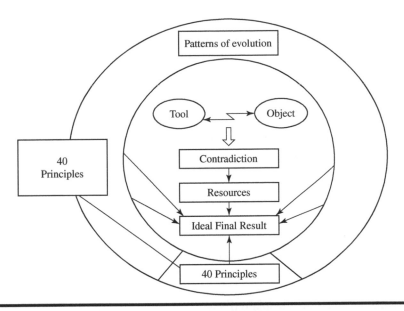

Figure 11.1 Forty Principles in the model for problem solving. The arrow from the box "40 Principles" to "Ideal Final Result" means that you can use principles as an independent tool. You can also use it for the development of the result after the analysis of contradictions and resources. Both shortcuts and longer ways are available in the model.

11.2 General Review of 40 Principles

Below is a list of all 40 Principles. Alternate names come from a variety of translations of the original Russian research.[4]

1. Segmentation (fragmentation)
2. Separation (taking out, extracting)
3. Local quality (local attributes)
4. Symmetry change (asymmetry)
5. Merging (consolidation)
6. Multifunctionality (universality)
7. Nested doll (nesting, *matryoshka*)
8. Weight compensation (antiweight, counterweight)
9. Preliminary counteraction (preliminary antiaction, prior counteraction)
10. Preliminary action (prior action, do it in advance)
11. Beforehand compensation (beforehand cushioning, cushion in advance)
12. Equipotentiality (bring things to the same level)
13. "The other way around" (do it in reverse, do it inversely)
14. Curvature increase (spheroidality, spheroidality curvature)
15. Dynamic parts (dynamicity, dynamization, dynamics)
16. Partial or excessive actions (do a little less)
17. Dimensionality change (another dimension)
18. Mechanical vibration
19. Periodic action
20. Continuity of useful action

21. Hurrying (skipping, rushing through)
22. "Blessing in disguise" (convert harm into benefit)
23. Feedback
24. Intermediary (mediator)
25. Self-service
26. Copying
27. Cheap disposables
28. Mechanical interaction substitution (use of fields)
29. Pneumatics and hydraulics
30. Flexible shells and thin films
31. Porous materials
32. Optical property changes (changing the color)
33. Homogeneity
34. Discarding and recovering
35. Parameter changes (transformation of properties)
36. Phase transitions
37. Thermal expansion
38. Strong oxidants (accelerated oxidation)
39. Inert atmosphere (inert environment)
40. Composite materials

The use of each of the principles is illustrated by examples from several different areas of technology and business. Many examples that were used earlier in the book are repeated here, to show how the 40 Principles can be used to develop solutions to those problems. Problems can be solved and systems improved in different ways, using one principle or using several together. In most solutions, more than one principle is used. When you find an interesting principle, look for other principles that can improve the idea. Often, one principle will give you a concept for a solution, but several may be necessary to get to a practical working solution. For example, when trying to simultaneously increase the durability of a gasket, as well as its ability to create a seal, several principles could be applied:

Principle 2—Separation (taking out): Create voids in a hard gasket to make it more flexible.
Principle 35—Parameter Changes: Change the shape of the gasket.
Principle 40—Composite Materials: Create a gasket with hard and soft characteristics. However, the ultimate solution to the gasket problem may well be the combination of a modified shape and composite material that contains voids.

To make it easier to read and remember, the list of principles is divided into groups of two to four. Each grouping is considered in a single section. The principles in some groups are naturally connected with each other; others are simply lists of different approaches. The groups are the following:

- Segmentation, separation (Principles 1, 2)
- Local quality, symmetry change, merging, multifunctionality (3–6)
- Nested doll, weight compensation (7, 8)
- Preliminary counteraction, preliminary action, beforehand compensation (9–11)
- Equipotentiality, "the other way around," curvature increase (12–14)

- Dynamic parts, partial or excessive actions, dimensionality change, mechanical vibration (15–18)
- Periodic action, continuity of useful action, hurrying (19–21)
- "Blessing in disguise," feedback, intermediary (22–24)
- Self-service, copying, cheap disposables, mechanic interaction substitution (25–28)
- Pneumatics and hydraulics, flexible shells and thin films, porous materials (29–31)
- Optical property changes, homogeneity, discarding and recovering (32–34)
- Parameter changes, phase transitions, thermal expansion (35–37)
- Strong oxidants, inert atmosphere, composite materials (38–40)

11.3 Segmentation, Separation (1, 2)

11.3.1 Principle 1

Segmentation. Fragmentation. Transition to microlevel. Divide an object or system into independent parts. Make an object easy to disassemble. Increase the degree of fragmentation or segmentation (see Figure 11.2).

Some examples are the following:

- The law of the transition to microlevel, considered in the previous chapter, is the result of the repeated application of the principle of segmentation. Segmentation is a very frequently used principle. It helps to "combine the incompatible" and meet contradictory requirements in many different problems. Recall the example of fire extinguishing. More effective firefighting (good) requires more water and causes water damage (bad). Water is necessary and, at the same time, there should be no water. Let us segment water to small droplets and then increase the degree of fragmentation to mist. Mist can suppress fire effectively, using a very small amount of water. In previous chapters, we processed the problem through many steps. It is possible to get the idea directly, using the single principle of segmentation.
- Stone washing is one of the technologies used to give denim the desired characteristics and appearance. Stones are, however, rather crude tools and special machinery is needed to handle them. A nice solution is to use enzymes instead of stones to get the same result. Enzymes are not small stones, but the concept of using molecules instead of large objects comes from the principle of segmentation.

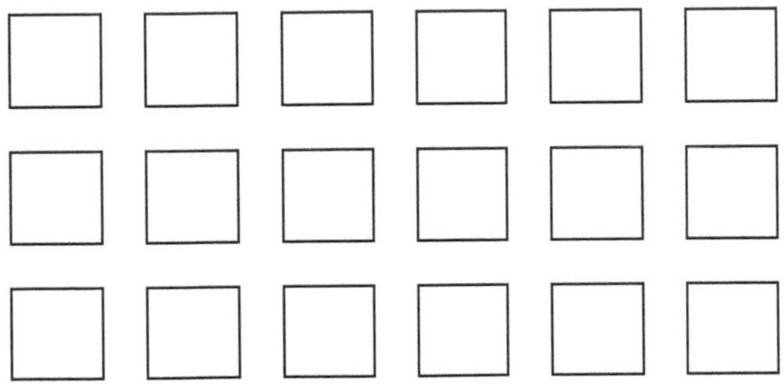

Figure 11.2   Principle 1—Segmentation. Divide a system into parts.

■ In previous chapters, we considered the noise problem of the lawnmower. It is useful to imagine different ways to segment or fragment the lawnmower, the muffler, and its environment. In the chapter on resources (Chapter 5), we speculated about using many small automatic minimowers instead of one big lawnmower. Sun-activated automatic lawnmowers, wandering over the lawn like sheep, have already been developed. In other applications, mini-robots are in use (e.g., for examination and cleaning of tubes). Perhaps someday the idea of the automatic lawnmower and mini-robots will be combined.

■ While waiting for the noiseless lawnmower, let us try to segment the muffler. The idea of using grass as the muffler (see earlier chapters) can be seen as an example of segmentation also. Obviously, this is not the only way of segmentation. Why not try replacing a single exhaust tube with many small ducts and a single muffler with many small ones? New materials and production technologies, such as casting of plastic, make it easy to get many components in one.

■ If we can segment the lawnmower, why not increase the segmentation of grass? Indeed, there are mulching lawnmowers that cut the grass into very small pieces. The benefit is that small pieces of grass will degrade and fertilize the soil, and there is no need to remove them from the lawn. This solves a different problem—waste removal instead of noise reduction—but it is common in TRIZ to find new opportunities for improvement.

■ Carrot cultivation is another example considered earlier. How can we use the segmentation principle here? We can imagine many minicarrots instead of one big plant. They can grow very densely, almost without soil, if they get water and necessary fertilizers (compare with hydroponics or cultivation in water). There are already gardening technologies for growing a greater number of little plants and getting a larger overall crop from the same area.

■ In Chapter 5 (resources), we also discussed the possible segmentation of the pin in the latching mechanism. The shape of a layered or filament pin can be controlled better than the shape of a solid component.

■ The segmentation principle has many applications in business. The "segmentation of the market" is a common practice and a commonly used term. Most big corporations have segmented themselves into business units or profit centers. A corporation should be small to be flexible and big to have enough resources for production and marketing. To be small and big at the same time, a huge company is divided into subsidiaries, profit centers, or other units working relatively independently. ABB and the Gore Corporation are exemplary—they create a new organization whenever an existing part of the company exceeds 150 people.

■ For the past 30 years, the use of teams has been one of the persistent themes in the workplace because small teams are flexible and can make decisions quickly.

■ A large job can be broken into many smaller jobs (called "a work breakdown structure" in project management). The JIT (just-in-time or Kanban) system uses the concept of segmentation to an extreme—it replaces the idea of mass production with the idea that the most efficient production system can produce a single unit just as easily as multiple units.

■ Entertainment examples, such as serialized novels and television movies, were considered previously. A new entertainment medium is emerging on the Internet, where a novel not only is serialized but also has several options in each chapter, so the reader can select the segments for a personalized book. This combines segmentation with Principle 3—Local Quality.

■ Segmentation is a good and often accessible way to use resources. The system can be divided into smaller parts. New components and substances are not needed.

More examples across industries are the following:

■ Make cupcakes instead of one large cake so that people can decorate according to their own tastes and a variety of flavors can be offered.
■ Use powdered welding metal instead of foil or rod to get better penetration of a joint.
■ Inject a drug in a finely powdered form.
■ Use Java applets. The predecessor of this example was the use of piping in UNIX to make large tasks possible by combining sequences of small tasks.
■ Paper is traditionally coated by a transferring application by a blade on a web. A new way is to spray coat in atomized form at high speed on both sides of the paper web.

11.3.2 Principle 2

Separation. Separate the only necessary part (or property) or remove an interfering part or property from an object or system (see Figure 11.3).

We usually need only a part of the system or some property or **attributes**.

For example:

■ We need light, not lighting devices. Today, many parts of the lighting system are located some distance away from the places lighted. Reflectors and fiber optics both can separate the mechanism, such as a lamp, from the point of use of the light. Fiber optics are used in tiny surgical instruments to provide light exactly where it is needed inside the surgical area without the bulk of a lamp.
■ Franchising separates the ownership of a local business, such as a restaurant or a printing shop, from the development of the concept and the systems that make it successful.
■ We do not need a vacuum cleaner as such, but rather cleaning capacity. A central vacuum cleaning system leaves only nozzles and a piece of tubing in the apartment. Noisy and dirty parts are located where they do not disturb inhabitants.
■ An electric lawnmower can work quite well if the lawn is not too large. The production of energy is removed from the lawn.
■ Put a noisy air compressor outside a building and pipe the compressed air to the place where it is needed—most medical and dental offices are built this way.
■ The example of fighting fire with mist illustrates the separation principle. Only small-diameter tubes are needed at the fire's location. The heavy part of the equipment is removed.

Does the paper producer need the paper-making machinery? Does an insurance company need mainframe computers and data storage? Only a few years ago, this question did not deserve attention. The insurance company outsources operation of its data center to a specialist company,

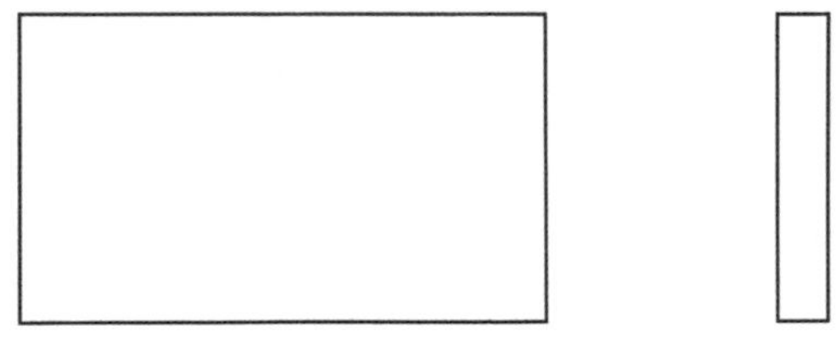

Figure 11.3   Principle 2—Separation. Separate a part or property from a system.

and the paper company outsources operation and maintenance of the mill. The ASP (application system provider) is a new business concept—many companies do not own their own software, but rather lease it as needed from a provider.

Do we need personal cars and bicycles? Until recently, the answer was "yes" for convenience. Today, new shared-use schemes are emerging. The user buys the right to use a car or a bicycle for a certain time. For cars, the credit card works as the key. In Portland, Oregon, the bicycle experiment requires the user to put the bicycle where the next person who needs one can take it. These experiments give us transportation capacity, individual routes and schedules, comfort, prestige, a certain lifestyle, and many other elements, but *separate* the use of the automobile from the ownership, care, and cost of a ton of metal and plastic. The proponents of these new schemes claim that a customer can actually have many cars by not owning any. Today, we can use a small vehicle in the city, tomorrow a big car for a long trip, and the day after tomorrow a limousine for prestige purposes. Costs can be cut because of the more intensive use of the capital invested in the automobiles.

Often, we have a contradiction between present and absent. Some awkward machinery or complex process should be present to get a needed **attribute** and the same machinery or process should be absent to save space, energy, and time. The separation principle may be a solution in problems of this sort.

More examples include the following:

■ Use a recording of a barking dog, without the dog, as a burglar alarm. Likewise, use the sound of birds in distress instead of a scarecrow.
■ Use a light pipe to separate the hot light source from the location where light is needed.
■ Outsource maintenance and operation services.

Exercise: Think of one example from your personal life or your business life for each of the principles in this section.

1.	Segmentation	
2.	Separation	

11.4 Local Quality, Symmetry Change, Merging, and Multifunctionality (3–6)

11.4.1 Principle 3

Local quality. Change an object's structure or an external environment (or external influence) so that the object will have different **attributes** or influences in different places or situations. Make each part of an object or system function in conditions most suitable for its operation. Make each part of an object fulfill a different and useful function (see Figure 11.4).

Often the object should have additional **attributes**, but the introduction of these **attributes** causes new problems or makes the system more complex and expensive. We should change the system, and we should not change the system.

Figure 11.4 Principle 3—Local quality. Change an object's structure.

It is easier to change the system locally. There are many examples in technology. Quenching and other treatments of the surface layer of metal components make the surface properties different from the bulk properties of the material. We use a different wrench for every nut because fixed-size wrenches are much stronger than adjustable wrenches. Specialized compartments in a lunch box for each type of food keep hot things hot, cold things cold, and make it safe and economical for workers and schoolchildren to carry their lunches with them.

In business, the segmentation of the market also illustrates the local quality principle. *Segmentation* is used to divide the market into small markets with specific attributes, and then *local quality* is used to treat each of those markets appropriately. To tailor its approach in the automatic washing machine market to the cultural preferences of each group, the Whirlpool Corporation has hired marketing people in India who speak 18 different languages.

Local quality applies to people, as well. Some are most effective working on their own, and others are most effective in teams. Intensive professional specialization is needed for certain skills, and a broad liberal arts background is required in other situations.

More examples include the following:

■ Precision farming using the correct amount of chemicals where needed
■ Pencil with eraser
■ Hammer with nail puller
■ Kids' areas in restaurants

11.4.2 Principle 4

Symmetry change. Change the shape of an object or system from symmetrical to asymmetrical. If an object is asymmetrical, increase its degree of asymmetry (see Figure 11.5).

Asymmetric paddles mix more effectively than symmetrical ones (both for concrete and for cake batter). Asymmetric scissors are easier to control than symmetric ones.

Increasing asymmetry is a way to use geometric resources. It is often easier to change the geometry than to introduce new substances or components.

Mass customization is a business strategy that corresponds to asymmetry—the product, service, or policies of a business are specifically designed for each customer and do not need to be the same as those provided to other customers.[5,6] This example could also be considered an application of the principle of local quality.

More examples include the following:

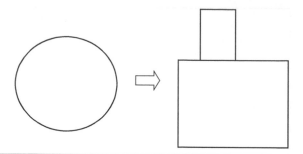

Figure 11.5 Principle 4—Symmetry change. Change the system from symmetrical to asymmetrical.

- Using astigmatic optics to merge colors
- Wider tires on the back of a truck than on the front
- Reducing the diameter of ductwork as the air moves further from the blower
- Budgeting for different departments individually rather than using a constant percentage increase or reduction for all departments

11.4.3 Principle 5

Merging. Bring closer together (or merge) identical or similar objects; assemble identical or similar parts to perform parallel operations. Make operations contiguous or parallel: bring them together in time (see Figure 11.6).

Examples of merging are the following:

- Integration in microelectronics.
- Telephone and computer networks.
- Swiss Army knife.
- Paper sheets constitute a book; books are merged into a library
- See examples of carrot seeding and of the handling of packages in previous chapters. Seeds can be placed precisely and quickly if fixed in position on the tape. Small packages can be moved easily if they are affixed to each other.
- Fragile and weak components, such as glass plates, can be made stronger without increasing weight by combining them into packages.
- An array of radio telescopes has greater resolving power than a single dish.
- An idea to get rid of traffic jams is to make vehicles travel in small "platoons" under computer control. A motorway lane could handle about 6000 vehicles an hour, instead of today's 2000.
- Flower garden.

Figure 11.6 Principle 5—Merging. Bring similar objects closer together.

If an object should be small *and* big and if there should be many *and* one, merging is often a solution.

In the section on segmentation (Principle 1), we discussed large corporations that are dividing themselves to get small at the same time. Small companies or individual entrepreneurs often have the opposite problem: how to get big while remaining small. Networking is perhaps the most popular solution. Others are chains of companies, franchising schemes, and conventional mergers. Segmentation and merging principles are often most effective if used together. Organizations are segmented and then the parts merge. (However, keep the principle of local quality in mind. The use of a single principle may lead to an ineffective or wrong solution.) Two or three principles together may work better than any one alone. In the 1970s, E. F. Schumacher launched the slogan "small is beautiful." Later, the slogan was forgotten because big can be just as beautiful.

The pattern of the transition to the macrolevel (Chapter 10) is the result of repeated applications of the principle of merging, and the pattern of expansion and convolution is the result of alternating merging with separation and with the next principle, multifunctionality.

11.4.4 Principle 6

Multifunctionality or universality. Make a part of an object or system perform multiple functions, and eliminate the need for other parts (see Figure 11.7). The number of parts and operations decreases and useful attributes and functions are retained. Some examples are the following:

■ ABB has developed an electric generator with high voltage. A conventional transformer is not needed because the generator can directly feed the electric network. In some new car designs, the flywheel, alternator, starter, and some other parts are combined into a single component. A single adjustable wrench for all nuts is another example.

■ People use the universality principle also. Cross-functional training makes people much less susceptible to layoffs because they have multiple skills instead of one skill. Compare this with the local quality principle.

More examples include

■ Team leader acts as recorder and timekeeper.
■ One-stop shopping—supermarkets sell insurance, banking services, fuel, newspapers, etc.
■ Handle of a toothbrush contains toothpaste.
■ Swiss Army knife.
■ Smartphone.
■ Bandage with infused antibiotics.

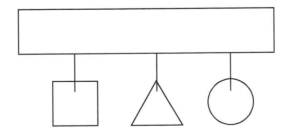

Figure 11.7   Principle 6—Multifunctionality. Make a system perform multiple functions.

Exercise: Think of one example from your personal life or your business life for each of the principles in this section.

3.	Local quality	
4.	Symmetry change	
5.	Merging	
6.	Multifunctionality	

11.5 Nested Doll and Weight Compensation (7, 8)

11.5.1 Principle 7

Nested doll. Place one object inside another; place each object, in turn, inside the other. Make one part pass through a cavity in the other. The name of this principle comes from the Russian folk art dolls or *matryoshkas,* in which a series of wooden dolls are nested one inside the other (see Figure 11.8).

Examples include the following:

- The double hull in oil tankers.
- Telescoping structures (umbrella handles, radio antennas, pointers).

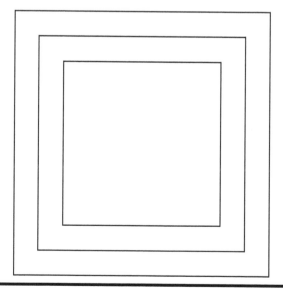

Figure 11.8   Principle 7—Nested doll. Place one object inside another.

- Business analogies: A special exhibit for one designer inside a boutique store, inside a big market.
- File structures in the Windows computer operating system (e.g., Chapter 11 is a file in the NewTRIZBook folder, which is a file in the My Documents folder, etc.).
- Measuring cups or spoons.
- Stuffing a turkey with sausage, stuffing the sausage with chestnuts, etc.
- Fractals.
- Elastic-connected concentric rings to create form-fitting seat.
- Metal stake made of concentric cylinders of harder and harder metals with hardest being at the core—natural wear keeps the stake "sharp."

11.5.2 *Principle 8*

Weight compensation. To compensate for the weight of an object or system, merge it with other objects that provide lift. To compensate for the weight of an object, make it interact with the environment (e.g., use aerodynamic, hydrodynamic, buoyancy, and other forces). See Figure 11.9.

Examples include the following:

- Air tanks in submarine vessels (modeled on air bladders in fish).
- Lifting bodies—the shape of the fuselage acts like a wing and generates lift. Used in both aircraft and ship design.
- Banners and signs cut so that the wind lifts them for display.
- Business analogies: Compensation for the heavy organization pyramid with project organization, process organization, temporary organization, and other less hierarchical systems "lift" the heavy structure.
- Helium balloon used to support advertising signs.
- Companies increase flagging sales by making connections with other rising products (e.g., movie tie-ins).

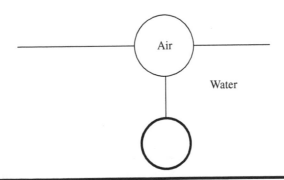

Figure 11.9 Principle 8—Weight compensation. To compensate for the weight of an object or system, merge it with other objects that provide lift.

Exercise: Think of one example from your personal life or your business life for each of the principles in this section.

7.	Nested doll	
8.	Weight compensation	

11.6 Preliminary Counteraction, Preliminary Action, and Beforehand Compensation (9–11)

11.6.1 Principle 9

Preliminary counteraction. If it will be necessary to do an action with both harmful and useful effects, this action should be replaced with antiactions (counteractions) to control the harmful effects. Create stresses in an object or system that will oppose known undesirable working stresses later on in time (see Figure 11.10).

Examples include the following:

- Use an electric heater to preheat the car engine before starting in the winter in northern regions. Damage to the engine from running with thickened cold oil is prevented, fuel is saved, and air pollution is decreased.
- Pre-tension rebar before pouring concrete for stronger structures.
- Changes and innovations usually meet resistance in an organization. Get the affected people involved so that they can participate in the planning of changes and do not feel threatened.
- Involve affected organizations in TRIZ solution generation.
- Use customer trials to launch high-risk new products (e.g., film companies shoot several endings to a movie and try them with different audiences before final selection).
- Focus on proactive instead of reactive maintenance.

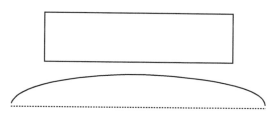

Figure 11.10 Principle 9—Preliminary counteraction. Introduce counteractions to control harmful effects.

11.6.2 *Principle 10*

Preliminary action. Perform, before it is needed, the required change of an object or system (either fully or partially). Prearrange objects so that they can come into action from the most convenient place and without losing time for their delivery (see Figure 11.11).

Examples include the following:

- Preliminary perforated packaging is easy to open.
- Precut parts for the building of wooden houses save work at the construction site.
- Workers arrange their workspace so that the most frequently used tools (physical, paper, or electronic) are the easiest to reach.
- Do market research, study possible futures, and build reserves for changes.
- Prepare solutions to problems that customers are not complaining about today, but may notice tomorrow: environment, moral code.
- Television demonstration chefs always have neat little dishes with all ingredients premeasured.
- Freeze "liquid" candy prior to dipping it in heated chocolate to create liquid-center treats.

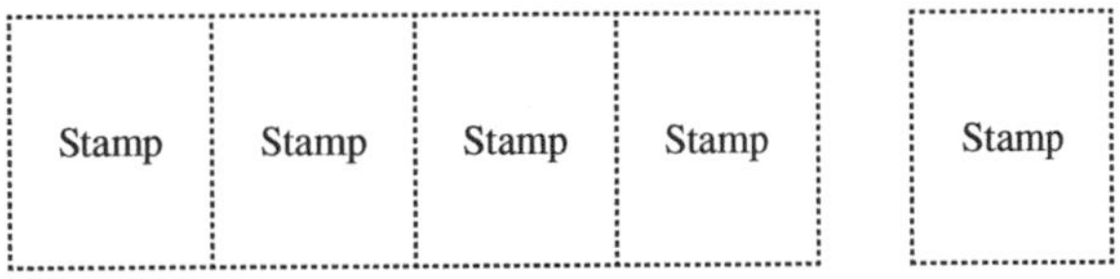

Stamp	Stamp	Stamp	Stamp	Stamp

Figure 11.11 Principle 10—Preliminary action. Perform, before it is needed, the required change of a system. Examples: Sheets of stamps are perforated before sale, to make separation easy. Food stores were arranged before the expedition to North and South Poles.

11.6.3 Principle 11

Beforehand compensation. Prepare emergency means beforehand to compensate for the relatively low reliability of an object or system over time (see Figure 11.12).

> Well-known technological examples are airbags in cars and pressure relief valves in boilers and chemical reactors. Airliner emergency oxygen masks drop from ceiling when needed. Sleeping pills now contain a small antidote in each tablet to prevent death by over consumption.

Nontechnical examples are posting instructions for possible emergency situations: fires, overdose or poisoning, environment problems, and preparation of equipment (first aid and rescue kits, fire extinguishers) where they may be needed.

The frequently asked questions (FAQs) sections of many websites are examples of Principle 11, where users are told how to help themselves solve problems that are known to exist in the system.

Principles 9 through 11 together constitute a group of time-related principles, either for preventing problems or for correcting them quickly.

Exercise: Think of one example from your personal life or your business life for each of the principles in this section.

9.	Preliminary counteraction	
10.	Preliminary action	
11.	Beforehand compensation	

Figure 11.12   Principle 11—Beforehand compensation. Prepare emergency means beforehand to compensate for the low reliability of a system.

11.7 Equipotentiality, "the Other Way Around," and Curvature Increase (12–14)

11.7.1 Principle 12

Equipotentiality. Change operating conditions to eliminate the need to work against a potential field (e.g., eliminate the need to raise or lower objects in a gravity field). See Figure 11.13.

Examples:

- Store all equipment, supplies, and materials at waist height so human does not have to reach or stoop to handle them.
- Use spring systems to lift sheets of wood to the right height so that workers can slide them into the machine for the next step in the process. The Bishamon Company makes these material-handling devices. When sold as productivity tools, they were a moderate success. When advertised as tools to prevent workers' back injuries, sales increased dramatically.
- Use grounding straps to bring people and objects to equal electrical potential to prevent harm from static electricity.
- Ascend slowly while scuba diving to bring internal body pressure to that of atmospheric pressure at the end of a dive.
- A business analogy might be a transition to a flatter organization with fewer hierarchical layers. One step in team formation is to bring all team members to the same level—eliminate hierarchical behaviors.

Figure 11.13 Principle 12—Equipotentiality. Eliminate the need to raise or lower objects or "walk uphill" some way.

11.7.2 Principle 13

The other way around. Invert the action(s) used to solve the problem (e.g., instead of cooling an object, heat it). Make movable parts (or the external environment) fixed and fixed parts movable. Turn the object (or process) upside down (see Figure 11.14).

Sometimes, this principle is applied very literally—turning machines upside down has solved many industrial problems. One core principle of effective design for manufacturing and assembly is to let gravity be your friend—always position parts so they will fall naturally into the desired place. Turn a carton so that the label can be applied on the top or turn an assembly so that screws can be inserted from the top. Examples include the following:

- Hang metal lath upside down so that metal shavings do not fall into mechanism.
- Instead of heating an object, cool its surroundings.
- Build products after customer places order.
- Extrude large pipes by moving extruder, not by pushing out extruded pipe.
- Place labels "upside down" on condiment containers so that they are stored with the opening down and the condiment is immediately ready for dispensing.
- Work at home instead of increasing travel time.
- Customers find their own answers in a consultant's database instead of having the consultant find the answer for them.
- Television and radio bring church services to people at home, instead of people going to the church.
- Hanging tomato gardens have been recently introduced where the plant is literally hung below its roots.

When considering an innovative principle, think also of the opposite idea. An industrial plant can be improved by making it flat. The need to raise and lower objects is eliminated (Principle 12). Sometimes, it may be better to build very many stories, according to Principle 17—Dimensionality Change. Any number of approaches, of course, can be combined.

Urban planning provides a mixed technological and social example, illustrating inverted approaches. The garden city movement is named for small cities with low buildings. Erecting skyscrapers is another

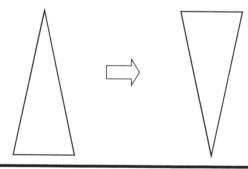

Figure 11.14   Principle 13—"The other way around." Turn the object or process "upside down."

approach. Le Corbusier's Radiant City combines both schemes. The center has a very high population density and 95% of the land remains free of construction.

More examples include the following:

- Benchmark against the worst instead of the best.
- Expansion instead of contraction during recession (this was one of the key ingredients to how Intel rapidly commanded 90%+ of the worldwide microprocessor market in a short 10-year period).
- Corporate unlearning—acquiring the ability to forget about the past when appropriate.
- Retinal projection into each eye (virtual reality monitor) instead of one big screen.

11.7.3 Principle 14

Curvature increase. Instead of using square, rectangular, cubical, or flat parts, surfaces, or forms, use curved or rounded ones; move from flat surfaces to spherical ones; from cube or parallel-piped shapes to ball-shaped structures. Use rollers, balls, spirals, or domes. Go from linear to rotary motion. Use centrifugal forces (see Figure 11.15).

The filament in incandescent lamps was initially straight. Efficiency improved when it was coiled. Photography was first done with flat plates of glass coated with sensitive emulsion. Cameras became portable when rolls of film were developed. The transition from flat surfaces to curved ones can be easily seen in cars and telephones. A mowing machine for agriculture started with a saw-like reciprocating edge. New machines have rotating blades similar to those in lawnmowers.

Corrugated forms often improve strength without increasing weight. Rotary motion frequently makes equipment simpler.

The principle of curvature increase is often paired with Principle 4—Symmetry Change (asymmetry). Components can be improved by making them more symmetric or more asymmetric. Curvature increase may increase or decrease symmetry.

Some nontechnical analogies can also be found. For example, increasing circulation of information benefits organizational function. Curved walls and streets make neighborhoods visually interesting (both in cities and inside large office buildings and schools).

More examples include the following:

- A curved shower curtain rod keeps the curtain from touching the person in the shower.
- Traffic circles instead of stop and go intersections.

Figure 11.15   Principle 14—Curvature increase. Move from rectangular forms to curved ones.

- Fuller's geodesic dome does away with bulky beams.
- Spherical casters are used instead of cylindrical wheels to move furniture.
- Hub and spoke organization structure.
- The Dyson vacuum cleaner spins dirt at high speed, forcing dust outward to the wall. The bag is not needed.
- Use iteration and design loops.

Exercise: Think of one example from your personal life or your business life for each of the principles in this section.

12.	Equipotentiality	
13.	"The other way around"	
14.	Curvature increase	

11.8 Dynamic Parts, Partial or Excessive Actions, Dimensionality Change, Mechanical Vibration (15–18)

11.8.1 Principle 15

Dynamic parts. Allow (or design) the characteristics of an object, external environment, process, or system to change to be optimal or to find an optimal operating condition. Divide an object or system into parts capable of movement relative to each other. If an object (or process or system) is rigid or inflexible, make it movable or adaptive (see Figure 11.16).

Repeated use of this principle and combination with the principle of segmentation results in the pattern of increasing interactions and the pattern of transition to a microlevel. Some steps in increasing dynamics are

- Monolith
- Shifted parameter monolith (different attributes at different locations)
- One hinge
- Many hinges
- Elastic system
- Powder system
- Liquid system
- Gas system
- A field (electromagnetic or otherwise) instead of a physical object or system

In Chapter 8, we presented an example of a lamp that was made more controllable by introducing joints. The penalty has been an increasing number of parts. The solution has been further improved by transition to elastic components. The single elastic component has many microlevel

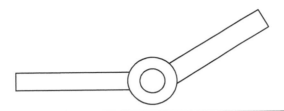

Figure 11.16   Principle 15—Dynamic parts. Make a system or process movable or adaptive.

parts, very many, very small joints. Here, Principle 1 (segmentation) helps the system to become more dynamic. Generally, if the improvement by one principle causes new difficulties, involve a different principle to solve the new problem.

Stiff and immovable structures are often replaced by more dynamic ones: flexible printed circuits and accumulator batteries in electronics, flexible and break-away light poles and signage on the roadsides, wings that change form in airplanes (through the use of flaps and slats on fixed-wing aircraft and through motion of the wings on fighter aircraft), and other dynamic structures. NASA has recently introduced highly efficient dynamic wing structures that change shape in nonsymmetrical fashion to take advantage of changing air pressures and wind currents.

A first step to making a building safe for earthquakes was to make it more rigid: thicker walls, for example. Later, to avoid impractical heavy structures, certain dynamics were added: the building now has bearings and shock absorbers that allow it to move a little. Strength is increased without extra weight. The automobile has gone through a similar evolution: first safety belts (passenger fixed more stiffly), then a dynamic part, the airbag, was added.

In business, flexibility—the capability to make changes when the environment changes—is often the difference between success and failure. Organizations are also evolving from rigid, unchanging structures to flexible ones. Ways to increase flexibility are segmentation, flatter organizations (see Principle 12), preparing changes before encountering a problem (Principles 9–11), discarding and recovering (Principle 34), and others.

Schools, too, have used the principle of dynamics as part of their improvement strategy. In many schools, students are no longer assigned to a fixed grade in which all 8-year-olds do third-grade studies together. Rather, the curriculum is flexible. One author's nephew recently was doing fifth-grade arithmetic, third-grade language studies, and a personal project to learn geography, all on the same day, in a program that was based on his abilities and interests.

More examples include the following:

- Software applications have user-configurable toolbars and interfaces.
- "Cafeteria" benefits allow employees to pick which types of insurance, health coverage, and such, they want.
- Traditional printed traffic signs are often replaced by electronic signs that vary in response to changes in weather and traffic conditions.
- In self-righting vessels, the ballast shifts, making the vessel right itself.
- Smart traffic lights help drivers to arrive when the lights are green and discourage frequent starts and stops.
- Flexible manufacturing systems (FMSs) are increasingly used in industry.
- To prevent hydroplaning, speed limits on highways can be adjusted, for example, according to the thickness of the water layer on the pavement.

11.8.2 Principle 16

Partial or excessive actions. If 100% of a goal is hard to achieve using a given solution method, the problem may be considerably easier to solve by using slightly less or slightly more of the same method (see Figure 11.17).

A classic example is to dip a brush in paint to acquire excess paint, then wipe the excess off. Similarly, attach a stencil to a surface to be painted, and then paint the whole thing. When the stencil is removed, the goal will be achieved, and the stencil will take the excess paint with it.

Perforated packages are easy to open. (Cut a little bit; do not cut the whole thing.)

Preparing sketches and concepts helps many writers get finished results more quickly.

If marketing cannot reach all possible customers, a solution may be to select the sub-group with the highest density of prospective buyers and concentrate efforts on them. Another solution is an excessive action: broadcast advertising will reach many people who are not potential buyers, but the target audience will be included in the group that is reached.

Figure 11.17   Principle 16—Partial or excessive actions. Make slightly less or slightly more.

11.8.3 Principle 17

Dimensionality change. Move an object or system in two- or three-dimensional space. Use a multi-story arrangement of objects instead of a single-story arrangement. Tilt or reorient the object, lay it on its side, use its other side (see Figure 11.18).

In our earlier example of the cultivation of carrots, plants are placed in rows—that is, one dimensionally. Square-foot gardening is a method developed for small gardens. Plants are placed in square blocks (a four-foot block subdivided into 16 one-foot squares), that is, two dimensionally. They are placed much closer than in large-scale agriculture. The yield is higher and weeds almost nonexistent because of the close spacing. Further, stack garden beds (like bunk beds) with plants needing less light on bottom floors.

Use of underground tunnels and buildings is increasing. At the same time, more and more high buildings, multilevel highways, and overpasses are built. In winter, Toronto becomes a mul-tistory city—people can travel more than five miles indoors by going from the basement of one building to enclosed walkways at the third floor of another, to the main lobby of the next. Also, see an earlier example of a garbage bin: vertical dimension is used to get space.

Sometimes, the additional dimension is invisible to the customer. Disney World in Florida pioneered the multidimensional concept now used in many amusement parks. A network of tunnels, workshops, dressing rooms, storerooms, and staff centers runs under the park. Characters are never seen in part of the park that does not match their role—instead, the worker vanishes from one area and uses the underground system to travel to the new area or to rest or remove or replace parts of a costume. This preserves the "magic" for visitors. Less glamorously, garbage is dumped into another system of tunnels, so visitors never see garbage being transported through the park.

More examples are

- Holograms as three-dimensional photographs.
- IMAX movies with three-dimensional effects.
- Curved shower curtain rods.
- Boeing's 747 jumbo jet has two stories for passenger seating.

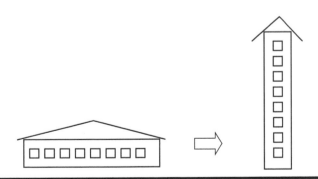

Figure 11.18   Principle 17—Dimensionality change. Move an object or system in two- or three-dimensional space.

11.8.4 *Principle 18*

Mechanical vibration. Cause an object or system to oscillate or vibrate. Increase the frequency of vibration. Use an object's resonant frequency. Use piezoelectric instead of mechanical vibrators. Use combined ultrasonic and electromagnetic field oscillation (see Figure 11.19).

Some technological examples include the following:

- Vibration instead of sound can be used to alert someone of an incoming call or message on a mobile telephone or pager.
- An object's resonant frequency is used for destruction of stones in the gallbladder or kidneys by ultrasound in a technique called "lithotripsy," which makes surgery unnecessary. This can also be seen as the use of segmentation because the stone breaks itself into very small pieces that the body then eliminates through its natural processes.
- Use resonance frequencies to enhance parts cleaning (ultrasonic cleaners).
- Use resonance frequencies to augment movement of material in a tube or on a conveyor.
- Wireless power transmission solution, ultrasonic sound waves generates electricity (see Section 7.7.3).
- Coordination is an analog of mechanical vibration similarly applicable in organizations. An example—working time and transportation schedules can be staggered and coordinated to decrease traffic congestion. Some people have also used vibration as a metaphor for putting a system into an excited state and then applied Principle 18 to various ways of exciting people to get coordinated action—examples range from cheerleaders or doing "the wave" at sporting events, playing music at political rallies, and so on.

Exercise: Think of one example from your personal life or your business life for each of the principles in this section.

15.	Dynamic parts	
16.	Partial or excessive actions	
17.	Dimensionality change	
18.	Mechanical vibration	

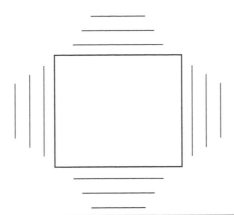

Figure 11.19   Principle 18—Mechanical vibration. Cause a system to vibrate.

11.9 Periodic Action, Continuity of Useful Action, and Hurrying (19–21)

11.9.1 Principle 19

Periodic action. Instead of continuous actions, use periodic or pulsating actions. If an action is already periodic, alter the periodic magnitude or frequency. Use pauses between impulses to perform a different action (see Figure 11.20).

Some examples are the following:

- In Whirlpool's washing machine, the pump pulsates. The company claims that the resulting wave effect removes dirt 40%–60% more effectively than a conventional agitator.
- Instead of a continuous light signal, a flashing light is often used for information, advertising, and warning.
- Researchers propose that taking naps in the middle of the day will increase the efficiency of intellectual work.
- Electrical energy for lighting and work is used most heavily in the daytime. Using financial incentives, power companies are attempting to get people to change the amplitude of the periodic action by moving power consumption to nighttime. In some warehouses, electric forklifts are programmed to recharge themselves between 2 a.m. and 5 a.m. because that is when power is least expensive.
- Online computer back-up systems usually run once a day at a preset time.
- Late model garden blowers have rapidly oscillating nozzles that create a pulsating air blast effect.
- Pauses in work can be used for training.

Figure 11.20   Principle 19—Periodic action. Use periodic or pulsating actions.

11.9.2 Principle 20

Continuity of useful action. Carry on work continuously; make all parts of an object or system work at full load all the time. Eliminate all idle or intermittent actions. Note that these last two principles contradict each other—if you eliminate all intermittent actions, you will not have any pauses to use. This just emphasizes that the various suggestions in each principle must be applied with common sense to a particular situation (see Figure 11.21).

The changing character of manufacturing shows considerable influence of both this principle and Principle 19 (periodic action). Lean and JIT manufacturing methods both emphasize small, customized production runs instead of long series. In this way, production is run continuously (Principle 20—Continuity of Useful Action) but across a range of product lines (Principle 19—Periodic Action).

More examples include the following:

- Continuous casting of steel and other metals
- Study during traveling
- Continuously variable automatic transmission (CVT)

Mechanical typewriters produced all lines in the same direction (depending on the language being written) with no writing during the time it took to return the carriage to the starting position. Electric typewriters with memory astonished the world when they showed the increased productivity of writing in both directions, thus eliminating the pauses between lines.

Figure 11.21   Principle 20—Continuity of useful action. Carry on work continuously.

11.9.3 Principle 21

Hurrying or skipping. Conduct a process or certain stages (e.g., destructive, harmful, or hazardous operations) at high speed (see Figure 11.22).

Examples include the following:

- During surgery, the longer a patient is anesthetized, the higher the risk of failure and future complications. Open-heart surgery that once took eight hours or more is now done in less than one hour, using combinations of new tools and methods. The injection gun is another example from health care: "Instead of allowing drugs to permeate gradually through the skin (such as with an ointment), new injection methods force them into the bloodstream at such high speeds that they pass straight through the outer layers of the skin. These pain-free injections do not use needles—they rely on gas pressure to inject the drug, which is prepared in a finely powdered form ... Since the guns do not penetrate the skin, they do not have to be discarded after a single use for reasons of hygiene."[7]
- A classic example of this principle is cutting plastic pipe very quickly. If you cut it slowly, heat from the cut region will propagate into the rest of the pipe, making it change shape.
- The traditional method for pasteurizing milk is to heat it to 72°C (161°F) for 15 seconds. Ultra-high-temperature pasteurization, in which milk is heated to 138°C (280°F), for only two seconds, increases the storage time of the product.
- In business, it may sometimes be more important to act quickly than to slowly produce perfect work. Thomas J. Watson, IBM founder, put it as follows: "If you want to succeed, double your failure rate." The importance of being first to market to establish a new standard has been emphasized in many studies of the e-business economy.

Periodic action (Principle 19), continuity of useful action (20), and hurrying (21) together compose a useful and practical set of tools. Keep them all in mind. The point is to apply the proper principles to the situation.

Agriculture is an illustrative example from technology. Seeding and fertilizing are typical periodic actions, following which the crop grows continuously. Hurrying is a good word to describe harvesting because the crop is at its perfect state of ripeness for a very short time.

Project management and personal time management are examples from business. Sometimes, hurrying is the most reasonable way (write a report or letter from the beginning to the end without pauses). If the job is big, it can be done only in parts (periodic action).

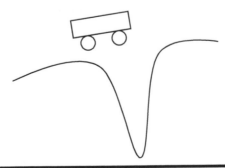

Figure 11.22   Principle 21—Hurrying. Conduct a process at high speed.

More examples include the following:

■ Fast cycle–full participation method: Involve the whole organization simultaneously in a major change such as a company reorganization.
■ Get the live lobster into the pot very quickly, holding the lid in your other hand, to prevent being splashed with boiling water.

Exercise: Think of one example from your personal life or your business life for each of the principles in this section.

19.	Periodic action	
20.	Continuity of useful action	
21.	Hurrying	

11.10 "Blessing in Disguise," Feedback, and Intermediary (22–24)

11.10.1 Principle 22

A *blessing in disguise* or "turn lemons into lemonade." Use harmful factors (particularly harmful effects of the environment or surroundings) to achieve a positive effect. Eliminate the primary harmful action by adding it to another harmful action to resolve the problem. Amplify a harmful factor to such a degree that it is no longer harmful (see Figure 11.23).

Some examples are

■ Electric charges that are usually harmful can be used for the control of process. Charges can cause fires and explosions, destroy electronic components, make materials stick, and do other harm, but, the same charges, present in most industrial processes, can give valuable information for the optimization of process. In coal firing, the measurement of electrical charges on coal dust is used to control burning, decrease harmful nitrogen oxide emissions, and improve efficiency.
■ Use harmful effect of an overcurrent situation to trigger the protection of circuitry (fuses, shunts, etc.).
■ Thermal expansion is often harmful and requires compensating devices, but, sometimes, it can be used to make a strong and reliable joint without fixing components. See Principle 37 (thermal expansion).
■ In an organization, complaints and destructive critique are negative "charges" that can be modified and used to bring about positive change.
■ Warrantee claims provide data that can be used toward product improvement and to prevent future claims.

- Singing the blues is an example also. The singer turns personal hardship into entertainment for others.
- Virus attacks in computer networks are never good, but each time the system survives an attack, information is generated that makes the system better protected from the next attack. This is similar to the way the body works—surviving an illness generates antibodies that protect the victim from the next incipient infection.
- In many situations, people have turned lemons into lemonade. The founder of TRIZ and author of *40 Principles*, Genrich Altshuller, began his work in the 1940s in Baku, in the former Soviet Union, at the time of Stalin. The government "rewarded" his activity by having him spend some years in work camps. There, Altshuller met many highly qualified experts who had been arrested during the great purges in the 1930s. He asked them to give classes and seminars. He established a "university" with one student and many professors. Thus, Altshuller obtained encyclopedic knowledge that he used to develop tools for problem solving. Altshuller's "University of One" is also an example of Principle 13—"The Other Way Around."

How can we amplify a harmful factor so that it becomes less harmful or becomes useful? Think about the harmful factor as a resource in a different process. Sulfur in coal, or other fuel, is a harmful component because the purification of the sulfur dioxide from exhaust gases requires complex technology and the unpurified gas is poisonous to people. However, if the sulfur content is increased, the production of sulfuric acid can become profitable. In the technology of flash smelting of copper (e.g., in Chapter 1), sulfur is used both as fuel and raw material for the production of sulfuric acid.

During the 1940s, one problem for combat aircraft was the potential explosion of gasoline when gasoline fumes mixed with air in partially empty tanks. Carrying an inert gas to fill the empty space in the tanks would require extra weight and complexity. In that case, another resource was used—the exhaust gases from the engine were produced on board the aircraft and had much less oxygen than ordinary air. The exhaust gases were pumped into the tank and prevented the explosive mixture from forming.

Setting backfires is a well-known technique for fighting forest fires. Controlled fires are set ahead of the fire being attacked to use up the fuel. When the fire reaches that area, it goes out.

More examples are

- Vaccination is a classic example of how to make harmful viruses to protect humans against themselves.
- Build up a tolerance to the allergen by exposure to an extract of the same allergen.
- Unhealthy salt, sodium chloride (NaCl), combined with bad-tasting potassium chloride (KCl) creates a healthy, good-tasting table salt.

Figure 11.23 Principle 22—"Blessing in disguise." Use harmful factors to achieve a positive effect.

11.10.2 Principle 23

Feedback. Introduce feedback to improve a process or action. If feedback is already used, change its magnitude or influence (see Figure 11.24).

A technical example—in a typical car, a driver makes observations and uses the steering mechanism, brakes, and other "actuators" to make necessary corrections. In new designs under development, the car has active driver-assistance systems with feedback. If the driver, for example, takes a curve too fast, the system controls braking and turns the steering wheel automatically.

The evolution of measurements and control is another example. Online measurements and online control are increased. Quality control in production is improved by introducing the immediate measurement and control during the production process, rather than inspection after production. In business, systems for getting feedback from customers are being continuously improved.

Feedback is a primary learning mechanism. Both babies and adults use it naturally, without thinking. People try something new. They examine the result. If it was successful, they do it again. If it was not successful, they modify it and then try it again. This applies to people learning TRIZ from this book, to a baby learning to walk, and to all other kinds of learning.

Modern business theory says that for optimum performance continuous real-time feedback must be provided to the employees as to how their efforts are affecting the organization's output. The author's hybrid vehicle has a similar feature. Real-time feedback as to instantaneous gas mileage provides insight as to how moment-to-moment decisions are affecting the miles per gallon. With this information, it is actually quite easy to insure higher mileage per gallon.

Another example is heart rate monitors for controlling intensity during exercise.

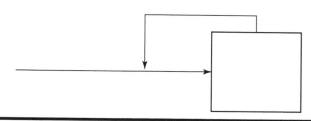

Figure 11.24   Principle 23—Feedback. Introduce feedback.

11.10.3 Principle 24

Intermediary. Use an intermediary carrier article or intermediary process. Merge one object temporarily with another that can be easily removed (see Figure 11.25).

Examples include the following:

■ Fixtures or jigs are used to position parts to make assembly easy. The assembled parts are removed and the jigs are used repeatedly. This concept is applied in the home as well as in the factory—a baking pan or a gelatin mold is an "intermediary" for shaping a dessert, and a pot holder is an intermediary for carrying a hot dish to the table without burning the server's hands.
■ A cable, or wireless system, is an intermediary between a computer and the Internet.
■ A power cord and plug are an intermediary between the electrical grid and an electrical appliance or lamp.
■ Ice can be used to hold small components in place temporarily if they will not be harmed by water when the ice melts. See also the earlier example of fixing carrot seeds on tape.
■ A neutral third party can be used as an intermediary during difficult negotiations. For sales promotion, an intermediary, who is seen by the customer as an impartial expert, can make recommendations.

Exercise: Think of one example from your personal life or your business life for each of the principles in this section.

22.	"Blessing in disguise"	
23.	Feedback	
24.	Intermediary	

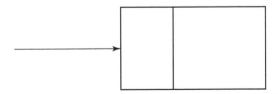

Figure 11.25 Principle 24—Intermediary. Use an intermediary carrier article or intermediary process.

11.11 Self-Service, Copying, Cheap Disposables, and Mechanical Interaction Substitution (25–28)

11.11.1 Principle 25

Self-service. Make an object or system serve itself by performing auxiliary helpful functions. Use resources, including energy and materials—especially those that were originally wasted—to enhance the system (see Figure 11.26).

Referring to Chapter 7 on functional analysis, remember that an auxiliary function is a function that affects components of the system, therefore self-service is the method of affecting the system (the tool) itself to improve or augment the operation of the system (e.g., use compressed exhaust gas to spin the turbine of a turbo charger on an automobile).

Some examples are

- In a tire that repairs itself, liquids are sprayed inside the tire. When the tire is punctured, the liquid fills the hole. When it contacts the outside air, it solidifies, forming a permanent repair. The leaking air performs an auxiliary function of moving the liquid sealant to the problematic hole.
- A classical example of self-service in business was presented in the beginning of the book: a self-service fast-food restaurant. Many electronic business ideas are based on including the customer and the customer's resources in the system as resources of the system—this includes everything from communities of interest and chat rooms to data exchanges such as Napster and Gnutella.
- Some search engines use the frequency of use of a Web site as the indicator of quality, so the more often a site is used, the higher it rates on their recommendation list. This is a combination of feedback (24) and self-service.
- Self-treatment and self-test: Patients themselves can perform some medical tests, such as the measurement of blood pressure or blood sugar or testing for fertility (then, later, testing for pregnancy) previously done only by medical personnel. In some cases, patients also adjust their treatment or behavior based on the test results (e.g., diabetes patients adjust their medication and diet).
- Water pressure on a hydrofoil lifts the hydrofoil out of the water, the hydrofoil, in turn, lifts the boat hull out of the water greatly reducing fluid friction on the boat.

Self-service is a way to use the object's resources. This principle illustrates the pattern of increasing ideality. What is more ideal than a system serving itself?

More examples are the following:

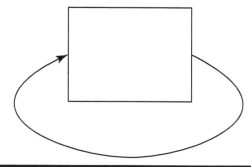

Figure 11.26   Principle 25—Self-service. Make a system serve itself.

- Halogen lamps regenerate the filament during use—evaporated material is redeposited.
- Lend out temporarily underutilized workers to other organizations. (Load-capacity balancing across companies creates a win–win situation where the worker [or player, in the case of sports teams] stays performance fit, the lender saves wages, and the lendee fills their skills shortage).
- A self-charging quartz watch is powered by the wearer's movement or incident light.
- A self-righting lifeboat can capsize and right itself again.
- Modern technology of melting steel scrap uses the energy of scrap itself; carbon and silicon are burned.

11.11.2 Principle 26

Copying. Instead of an unavailable, expensive, or fragile object, use simpler, inexpensive copies. Replace an object, system, or process with optical copies. If visible optical copies are already used, change the wavelength to infrared or ultraviolet (see Figure 11.27).

Some examples are the following:

- Make measurements from an image instead of directly. This includes a wide spectrum of technologies, from satellite photographs of farm and timber resources to ultrasonic images of a fetus in the womb.
- Use a simulation instead of the object. This applies to many business processes as well as to products and services.
- Use prototypes for testing new systems so that any harm is detected early.
- Use virtual prototypes instead of physical ones.
- Use videoconferencing instead of travel.
- Use virtual reality to test new processes or to train people to do work in difficult situations. Surgeons now test new operating procedures on virtual patients, and automobile assembly workers practice new procedures in virtual factories.

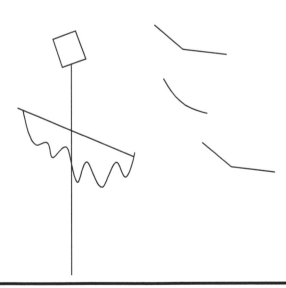

Figure 11.27 Principle 26—Copying. Use inexpensive copies. For example, a scarecrow instead of a person.

■ Scan rare historic books, documents, and such so they can be made accessible to all, while the original remains protected.
■ Use telepresence instead of fully independent robots.

Fake furs and leathers are also examples of the copying principle. Artificial grass might be an acceptable alternative in some places. No need to mow. Ultraviolet light shows certain kinds of skin lesions better than visible light; dyes that are sensitive to ultraviolet are used to find cracks in metal parts. Infrared images capture heat zones—this is the basis of most night-vision systems.

11.11.3 Principle 27

Cheap disposables. Replace an expensive object with multiple inexpensive objects, compromising certain qualities (such as service life, for instance; see Figure 11.28).

H&M (Hennes & Mauritz AB), a multinational clothing-retail company from Sweden, sells trendy fashionable clothing at inexpensive prices. The clothing only lasts a couple of years but serves the changing tastes of its consumers well.

Other examples include the following:

■ Disposable paper and plastic tableware
■ Disposable surgical instruments
■ Disposable protective clothing
■ Replaceable teeth on the shovel of a piece of earth moving equipment

Figure 11.28 Principle 27—Cheap disposables. Replace an expensive object with multiple inexpensive objects.

11.11.4 *Principle 28*

Mechanical interaction substitution. Replace a mechanical method with a sensory (optical, acoustic, taste, or smell) method. Use electric, magnetic, and electromagnetic fields to interact with the object. Change from static to movable fields to those having structure. Use fields in conjunction with field-activated (e.g., ferromagnetic) particles (see Figure 11.29).

Some examples include the following:

- The best-known example of the use of a smell as a warning is the incorporation of bad odors into natural gas to warn users when the system has a leak.
- The JIT manufacturing systems use Kanban cards or objects such as portable bins to indicate visibly when supplies are needed.
- See Chapter 10 for the discussion of the pattern of evolution called "the increase of interactions." The history of technology is full of examples in which the mechanical means of doing something is first supplemented by an electrical system and then replaced by an electrical or electronic system. Automobile steering systems are mainly mechanical, but control by wire (already used extensively in aviation) is intensively studied in the automotive industry. In telecommunications, infrared and radio waves and other wireless technologies are increasingly used.
- In the case of automatic solar lawnmowers, something must prevent them from escaping to a neighbor's lawn. One solution is a sensor picking up a signal from a low-voltage (also solar-powered) cable buried out of sight.
- In communication and business, we also clearly see the increase of new interactions. When human society began, all communication was face-to-face, which has since been augmented by writing, telegraph, telephone, fax, e-mail, videoconferencing, and other means.
- Transition to more easily controllable interactions is often associated with transition to the microlevel or segmentation. In inkjet printers, ink particles are controlled by thermal or electromagnetic fields. In video displays, text and figures are produced, changed, and removed using electromagnetic fields to control microparticles or molecules—for example, many flat-panel computer displays use liquid crystals, in which the image depends on the reflection of light from the molecules and the reflection is modified by changing the orientation of the molecule.

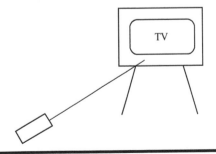

Figure 11.29 Principle 28—Mechanical interaction substitution. Use acoustic and optical interactions, taste, and smell. Use electromagnetic fields.

- Maglev vehicles use magnetic fields to levitate above a guideway.
- Facial recognition systems instead of human driven identification for access to secured areas.
- Magnetic strip cards and smart cards are used instead of paper cash and checks.
- A ring laser gyroscope, unlike the old mechanical gyroscope, has no moving parts.
- CD devices with laser beam superseded old record players with the diamond-tipped pick-up arm. In turn, digitally stored, moved and played music has superseded the older compact disc technology.
- Have retail customers enter data by means of a touch screen, instead of filling out a form.

Exercise: Think of one example from your personal life or your business life for each of the principles in this section.

25.	Self-service	
26.	Copying	
27.	Cheap disposables	
28.	Mechanical interaction substitution	

11.12 Pneumatics and Hydraulics, Flexible Shells and Thin Films, and Porous Materials (29–31)

11.12.1 Principle 29

Pneumatics and hydraulics. Use gas or liquid as parts of an object or system instead of solid parts (e.g., inflatable, filled with liquids, air cushion, hydrostatic, hydroreactive). See Figure 11.30.
Some examples are

- One way to transition to the microlevel is through the use of pneumatics and hydraulics. Examples of pneumatics are inflatable houses or air houses (stores, exhibition pavilions, sports halls, and the like), inflatable boats, and the moving of heavy components with an air cushion.
- Inflatable components allow the designer to decrease weight without losing strength. Air cushions enable movement of heavy objects using very little energy—for instance, they are used instead of mechanical jacks to raise aircraft that are on soft surfaces such as grass or mud.

Boat ⇨ Inflatable boat

Figure 11.30   Principle 29—Pneumatics and hydraulics. Use gas and liquid instead of solid parts.

- Similarly, hydraulic equipment can exert more force than simple mechanical systems of the same size. Replacing a physical cutting blade with a water jet illustrates the hydraulics principle of replacing solid objects with liquids.
- Business analogies use the idea of fluidity replacing rigid structures. Insurance companies now have "fluid" systems in which customers modify policies to suit their own needs. There is a lot of overlap with Principle 15 (dynamic parts).
- In 1997, NASA used air bags, reducing the use of expensive rockets, for the landing of Mars Pathfinder.
- Hydraulic brakes replaced pure mechanical systems many years ago.

11.12.2 Principle 30

Flexible shells and thin films. Use flexible shells and thin films instead of three-dimensional structures. Isolate an object or system from the external environment using flexible shells and thin films (see Figure 11.31).

Examples include the following:

- Coatings were historically spread on paper by means of a blade. One improvement is to transfer coating to the paper in the form of thin film. The paper is more evenly coated.
- Medication in pill form is dispensed over time by layering thin films that dissolve at different rates around small portions of the medicine. Patients can take one pill and the medicine will be released into their system over time.
- Some materials, such as epoxies, have two components that must be kept separate until shortly before use. Packages have premeasured portions that are separated by a thin film that the user can easily break to mix the product, making it convenient to use. This also has an element of Principle 10.
- Heavy glass bottles for drinks are often replaced by cans made from thin metal (aluminum) or thin plastic material. Hydraulics is also used—the pressure of the drink makes the can stiff. Weight is reduced without deterioration of strength.
- The sapphire-tipped temperature sensors from the Intel case study (see Section 12.3) were protected primarily by way of flexible shells and thin films.

Figure 11.31   Principle 30—Flexible shells and thin films. Use flexible shells and thin films instead of three-dimensional structures.

11.12.3 Principle 31

Porous materials. Make an object porous or add porous elements (inserts, coatings, etc.). If an object is already porous, use the pores to introduce a useful substance or function (see Figure 11.32).
Some examples are the following:

- A ceramic filter made from hydrophilic porous sintered material simplifies and improves vacuum systems. Water that passes through the filter seals the filter at the same time. The air leak of the conventional vacuum filter is eliminated, and the consumption of energy decreased. Similarly, porous ceramics are used in many hazardous-waste-cleanup systems—the pores can be filled with materials that bind the hazardous substances, and the ceramic particles make it much easier to distribute and remove these materials.
- High-tech microfibers are now well known. Small pores prevent water from passing through, but allow moisture to evaporate. New plasters, or wound dressings, are less known but not less exciting. They are covered by semipermeable film. Water, dirt, and bacteria are kept out, but excess moisture can evaporate through.
- Memory foam mattresses utilize porous foam block.
- Porous asphalt allows alternatives for managing storm water.
- Business analogy: Make the organization "porous"—make it easy for information to flow through the system. Make it easy for customers to penetrate the organization to find what they need.

Exercise: Think of one example from your personal life or your business life for each of the principles in this section.

29.	Pneumatics and hydraulics	
30.	Flexible shells and thin films	
31.	Porous materials	

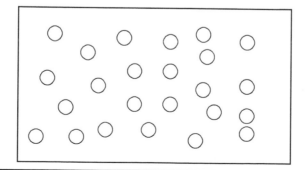

Figure 11.32 Principle 31—Porous materials. Make an object porous.

11.13 Optical Property Changes, Homogeneity, and Discarding and Recovering (32–34)

11.13.1 Principle 32

Optical property changes. Change the color or transparency of an object or its external environment (see Figure 11.33). Principle 32 can also be used to trigger thoughts of changes in wavelength or frequency of an event.

Some examples are the following:

- Enzymes producing light can be used for detecting impurities in food. Toys for preschool children are packed in transparent packages. Sunglasses that change the amount of light blocked, depending on the brightness of the environment, display both Principles 32 and 25 (self-service).
- "Transparency" is both a physical and a business term. Changing the transparency—increasing or decreasing it—is one important and often cheap way to improve business. Children want to see the toy; adults want to see that products and the whole production process are safe and ecological. Here, transparency can make the difference between success and failure.
- Business school case studies often cite Johnson & Johnson for its exemplary "transparent" behavior as a form of crisis management. When there was a tampering problem with its Tylenol product, the company immediately recalled the product and gave the public full information on its actions. Both the speed and the full disclosure of the situation are credited with the product's recovery and the company's retention of its excellent reputation.
- Porous materials and optical property changes compose a pair of principles that are frequently used together. Porous materials are often semitransparent. A business analogy is the firewall in a computer system. The wall should be transparent for legitimate users and should, at the same time, be impermeable to those who might try to steal essential information.
- Use of lighting effects to change mood in a room or office.
- Highlighter pens.
- Creation of corporate colors—creating a strong brand image through the use of bespoke colors—BP green, British Telecom red phone boxes, Ford blue, and so on.
- "Tell the truth" tape turning a darker gray when food is spoiled.
- Clear dirt canisters on vacuums allow the user to see when the canister needs emptying.
- Fuel gauges on cars and trucks turn color to indicate when fuel is low and then again when the fuel is almost all gone.

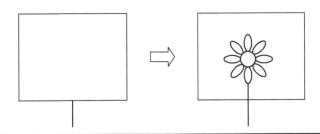

Figure 11.33   Principle 32—Optical property changes. Change the color or transparency of an object.

11.13.2 Principle 33

Homogeneity. Make objects that interact out of the same material (or material with identical properties). See Figure 11.34.

In the plastic industry, beads of polystyrene are frequently shipped in bags made of polystyrene. That way, the company that melts the beads can put the whole package, including the bags, into its processing equipment. This saves time—no need to open the bags—and eliminates the need to store and dispose of or recycle the bags.

In food, many products use the idea that the wrapper should have the same properties (be edible) as the contents. Ice-cream cones, tacos, and spring rolls are good examples.

The new wound dressings that we mentioned earlier act like a second skin. They keep wounds moist. Wounds themselves are moist and heal better in a moist environment. This innovation has an instructive history. From ancient Egypt to Rome and through the Middle Ages, wounds were treated by keeping them moist. Then this knowledge was forgotten and rediscovered only in the 1960s.

In business, this principle can be used by analogy. People may be more ready to buy things that remind them of familiar products than those that look very different. We are all familiar with movie sequels. The Harry Potter series contained no less than eight films, and, prior, *The Godfather* was followed by *The Godfather II* and *The Godfather III* (and the *Rambo* series, the *Star Wars* series and the *Jurassic Park* series, etc.) Of course, the principle should be used together with the opposite idea. To make the movie sequel, you must first have an original movie, different from all others.

More examples include the following:

■ Co-located project teams
■ Bioglass that bonds with natural bone
■ Boeing "working together" teams—bringing customers and suppliers into the design loop.
■ Light bulbs replicating the balanced spectrum of more comfortable natural sunlight.
■ Instead of electric light, natural sunlight carried with light guides.
■ Asphalt-infused tires to reduce tread wear
■ Spatula made of nonstick coating materials for use with nonstick coated pans.

Figure 11.34   Principle 33—Homogeneity. Make objects that interact out of the same material.

11.13.3 Principle 34

Discarding and recovering. Make portions of an object that have fulfilled their functions go away (discard by dissolving, evaporating, etc.) or modify them directly during operation. Conversely, restore consumable parts of an object directly in operation (see Figure 11.35).

Some examples include the following:

- Biodegradable materials in medicine: Polylactides are used to make dissolvable screws and pins. They can replace titanium screws used by surgeons to mend broken bones. The second operation for the removal of screws is not needed. Another example is the tread of a tire that restores the edges of the tread blocks as the tire wears.
- Many recovery processes have been used in technology for a long time. Chemicals used for pulping wood are recovered in a recovery boiler. Environmental requirements will make recovery more important. For example, water polluted in some process is more and more often purified and recycled back to the process.
- In business, the project organization is a good example of discarding and recovering. A good project should have an end. The organization will then be dissolved. The members can use their skills again in new projects. In all work, knowledge and skills are updated and improved directly during work and by retraining.
- Closed system sand blasting equipment recovers the sand and continually reuses it.
- A halogen bulb uses the discarding and recovering process to "recondition" its filament as it is "used up."

Exercise: Think of one example from your personal life or your business life for each of the principles in this section.

32.	Optical property changes	
33.	Homogeneity	
34.	Discarding and recovering	

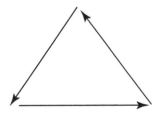

Figure 11.35   Principle 34—Discarding and recovering. Make portions of an object that have fulfilled their functions go away. Restore consumable parts.

11.14 Parameter Changes, Phase Transitions, and Thermal Expansion (35–37)

11.14.1 Principle 35

Parameter changes. Change an object's physical state (i.e., to a gas, liquid, or solid). Change the concentration or consistency. Change the degree of flexibility. Change the temperature (see Figure 11.36).

Some examples include the following:

- The new medical plaster described earlier also illustrates a parameter change. A traditional dry plaster is replaced by a moist one. The principles of parameter change, homogeneity, and use of porous materials are combined in this technology.
- Transport gaseous nitrogen or oxygen as liquids; similarly, transport natural gas and propane as liquids.
- Change the melting temperature of chocolate so that it does not melt when it is exposed to high temperatures.
- Powdered paints can be used instead of liquid ones. Powder paint combines the performance of modern latex and silicon-based emulsions with the convenience of the powder. Powder can be easily transported and stored. To use, just add water.
- In business situations, a parameter change is frequently realized as a policy change. In the past decade, many companies have increased the flexibility of employee benefit programs—instead of having one standard program, employees can design a mix of medical and life insurance, pension plans, and so on. Likewise, mass customization systems let customers have much more flexibility in designing products that exactly fit their needs (also a demonstration of Principle 15—Dynamic Parts).
- Use flexible car bodies that can withstand some impacts without requiring repair.
- Nonstick pans employee a parameter change of their surface to reduce the adhesion of the materials baked within them.

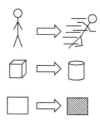

Figure 11.36   Principle 35—Parameter changes.

11.14.2 Principle 36

Phase transitions. Use phenomena occurring during phase transitions (e.g., volume changes, loss or absorption of heat). The most common of the many kinds of phase transitions include solid–liquid–gas–plasma, paramagnetic–ferromagnetic, and normal conductor–superconductor, but many useful phenomena are associated with more exotic transitions as well, such as solid–solid crystallographic changes, superfluidity, antiferromagnetism, and such (see Figure 11.37).

Examples include the following:

- Blasting with solid carbon dioxide can clean surfaces. Impurities will freeze immediately, contract, and loosen easily. The example also illustrates parameter changes (35), thermal expansion (37), and hurrying (21). Impurities are frozen so quickly that the cleaned material does not suffer from thermal expansion. The carbon dioxide then sublimes (harmlessly turns into gas), so there is no cleanup needed.
- Heat pipes are well-known examples of using the phenomena associated with phase transitions. The heat given up and absorbed as the fluid in the pipe transitions from liquid to gas can be used either as a heater or air conditioner, depending on how the system is arranged.
- "Muscle wire" is a form of nickel-titanium alloy that is used in robot systems and in orthodontics. Small electric currents heat the wire, which goes through a crystallographic phase change that changes its length.
- When businesses make structural changes (mergers, acquisitions, or internal changes) the accompanying phenomena are analogous to heat in a phase change—there is lots of confusion. Constructive ways to use this period of disruption include finding new means of aligning business systems with new strategies, forming new alliances with customers or suppliers, and getting rid of obsolete practices.
- Thermal inkjet printers use vaporization to force ink out of the nozzle.
- Business analogies of phase transitions may also be required at different stages—conception, birth, development, maturity, retirement—of a project.

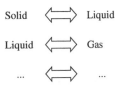

Figure 11.37 Principle 36—Phase transitions.

11.14.3 Principle 37

Thermal expansion. Use thermal expansion (or contraction) of materials. If thermal expansion is being used, choose multiple materials with different coefficients of thermal expansion (see Figure 11.38).

Thermal expansion can be used to position and fit components such as a valve in an engine. A component is cooled in liquid nitrogen, contracts, is installed, then expands, and fixes itself in position. A kitchen example is loosening a tight metal lid on a glass jar by running it under hot water. The metal lid expands more than the glass jar, so it is easier to remove the lid. Older thermostats used two pieces of different metals with different coefficients of expansion in order to change the position of the contact to turn on or shut off the current. The two metals were sandwiched together so as the temperature changed, the metal sandwich would either bend or straighten thus opening or closing the electrical contact.

Exercise: Think of one example from your personal life or your business life for each of the principles in this section.

35.	Parameter changes	
36.	Phase transitions	
37.	Thermal expansion	

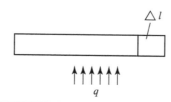

Figure 11.38   Principle 37—Thermal expansion. Use thermal expansion.

11.15 Strong Oxidants, Inert Atmosphere, and Composite Materials (38–40)

11.15.1 Principle 38

Strong oxidants. Replace common air with oxygen-enriched air. Replace enriched air with pure oxygen. Expose air or oxygen to ionizing radiation. Use ionized oxygen. Replace ozonized air (ionized oxygen) with ozone (see Figure 11.39).

Oxygen is used in the bleaching of pulp (for paper production). There are ideas and experiments to use ozone for bleaching.

Jules Verne wrote a novel called *Dr. Ox's Experiment.* In Quiquendone, an imaginary town, life is very slow. Then Dr. Ox injected a gas into the town and everything started happening very fast. People got exceptionally lively and passionate. Perhaps, we also sometimes need something like Dr. Ox's gas to charge the mental atmosphere with positive effects.

Examples include the following:

- Using simulations and games instead of lecture-style training.
- Scuba diving with Nitrox or other gas mixtures to extend endurance.
- Treating wounds in a higher-pressure oxygen environment to kill anaerobic bacteria.
- Metal active gas (MAG) welding.
- Using ozone to kill viruses, bacteria, and fungi in wounds. It does a better job of sterilizing than other agents, and it is much faster since the doctor does not need to analyze what type of organism is causing the infection.

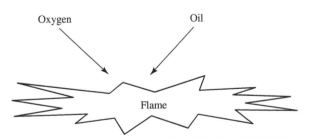

Figure 11.39   Principle 38—Strong oxidants. Make the process more active.

11.15.2 Principle 39

Inert atmosphere. Replace a normal environment with an inert one. Add neutral parts or inert additives to an object or system (see Figure 11.40).

Some examples are the following:

- Inert gases (such as carbon dioxide or argon) are used in welding to prevent oxidation of the material at the weld.
- Inert materials are added to detergents to make them easier for consumers to measure. (Did you ever wonder why the box says "97% inert materials"?)

Strong oxidants and inert atmosphere can be considered as a pair of principles. Using them together frequently gives good results. Oxygen is used to generate needed energy and inert gases are used to prevent undesired oxidation. An inventor proposed using a welding device as a fire extinguisher. A device is provided with a simple system that can blast inert gas with high pressure.

A social analogy of inert atmosphere may be indifference and neutrality. In business, negative situations mostly need cooling. Ignore or neutralize negative and destructive actions (if you cannot turn them positive; see Principle 22). Use neutral arbitrators. Here, too, the point is to use the correct approach at the right time.

Another example is metal inert gas (MIG) welding.

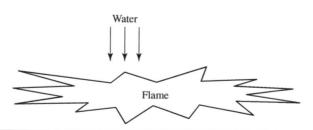

Figure 11.40   Principle 39—Inert atmosphere. Make process slower and more passive.

11.15.3 Principle 40

Composite materials. Change from uniform to composite (multiple) materials and systems (see Figure 11.41).

Composite materials are the combination of two, or more, materials with different characteristics such as seen in concrete, metal composites, and fiber-reinforced polymers. As an example, carbon fiber materials are now being used for automobile and aircraft skins as the materials are very strong yet light.

Some examples are the following:

- Rubber reinforced with woven cords, reinforced concrete, and glass-fiber-reinforced plastics are typical technology examples.
- The use of *nothing* (air or vacuum) as one of the elements of a composite is very typical of TRIZ—the resource of "nothing" is always available. Examples include honeycomb materials (egg crates, aircraft structures), hollow systems (golf clubs, bones), and sponge materials (packaging materials, scuba diving suits.) These combine Principle 31 (use of porous materials) with Principle 40 (use of composite materials).
- In business, we can speak of composite structures as well. Multidisciplinary project teams are often more effective than groups representing experts from one field. Multimedia presentations often do better in marketing, teaching, learning, and entertainment than single-medium performances. Other examples are less tangible but not at all less important. Fanatical commitment to cleanliness is one famous aspect of McDonald's. Consistent preparation of food is another major commitment. These are two principles or values or fibers that tie together a loose organization.

The principle of composite materials or, more generally, composite systems, is a good conclusion for this section on using inventive principles. If you have a system, you can improve the result by combining it with another system. Innovative principles are also systems. Composite principles often do better than single ones.

More examples are

- Combined high-risk/low-risk investment strategy.
- Flammable polyurethane coated with fire-resistant Kevlar (e.g., in airplane seat cushions).

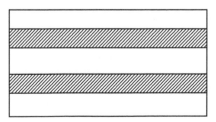

Figure 11.41   Principle 40—Composite materials. Change from uniform to composite systems.

Exercise: Think of one example from your personal life or your business life for each of the principles in this section.

38.	Strong oxidants	
39.	Inert atmosphere	
40.	Composite materials	

You can make the 40 Principles a more effective toolkit and tailor them to your needs and purposes by adding your own examples. Some people find they get the most creative stimulus from examples outside their own field, and others prefer to start with examples that are close to their own field. If you share your examples with others, you will contribute to the continuous development of problem-solving tools.

11.16 How to Select the Most Suitable Principles

The easiest way to find a useful principle is simply to browse the list in a relaxed way. Knowledge of the system and the constraints in the problem will help you decide which principles are suitable.

Let us consider five earlier examples: reducing lawnmower noise, making it easy to grow carrots, improving the pin-type latch, making training more effective, and fighting fire with less water. It is easy to see the applicability of certain principles to each of the examples.

Table 11.1 reflects typical results obtained by quickly browsing the list of principles. Closer examination reveals more principles—perhaps more suitable ones. These results can be improved using some guidelines for the selection of the most helpful principles.

11.16.1 Tradeoffs

Principles can be selected by tradeoffs (considered in Chapter 3). When the tradeoff is formulated, it can be used to eliminate unsuitable principles. Principles that negate the tradeoff are those that should be selected. For example, when noise absorption improves, the size of the muffler and the number of parts increase. Consider the multifunctionality or universality principle (6). If the increasing size and part number are the problem, it is wise to check whether some other resource, for example the casing or the grass, can take the job of the muffler.

Table 11.1 Examples of Problems and Principles that Can Be Used to Resolve Them

Problems	Principles
Reducing lawnmower noise	Porous materials (31)
Making it easy to grow carrotsImproving pin-type latch	Merging (5), multifunctionality (6) Segmentation (1), dynamic parts (15)
Making training more effective	Segmentation (1), composite systems (40)
Firefighting with less water	Segmentation (1)

The more precisely carrot seeds are planted, the slower the speed of planting. There is a process—planting. Second, there are many seeds, which are difficult to handle. Preliminary action (10) and the intermediary principle (24) seem natural. Both help remove the contradiction by suggesting that the carrot seeds could be arranged on the tape before planting.

If a pin is made easy to lock and open, wear increases. The same component should meet contradictory requirements and have incompatible properties. The segmentation principle (1) may be the first approximation because the parts and the whole can have opposite properties. For example, if the pin is segmented to two parts so that one part is a small wedge, the latch may be easier to open. Use of asymmetry (4) will also help—the pin can have different dimensions at the top, middle, and bottom because each area has different requirements.

As training gets more thorough, it requires more time. These are contradictory requirements for the same activity. Segmentation (1), multifunctionality (6), preliminary action (10), and composite materials (40) may be the first principles that come to mind. Lengthy training can be broken down into smaller parts that can be incorporated into work and perhaps prepared beforehand.

The more water that is used for fighting fires, the more equipment is needed and the more damage will occur. Consider segmentation (1). Again, the same component should meet contradictory requirements and have incompatible properties. First, we can try to fragment water in some way.

Additional cues for finding principles for resolving tradeoffs are summarized in the Contradiction Matrix. The matrix is a big table. The rows are labeled with typical features that are improving within the initial contradiction, and the columns are labeled with the features that get worse within the initial contradiction. The cell at the intersection of the row and column has the numbers of the principles that were historically used most frequently to solve this particular tradeoff (see Table 11.2 for a sampling of the full matrix).

Typical features of the matrix are presented in Table 11.3 (and Reference 8) and in a modification of the full matrix, which appears as an appendix at the end of this book. Altshuller developed it to help organize the principles he created from his analysis of the patent database. The original matrix is also available on the Internet.[10] In our version, the numbers in the cells are the same, but cells that were empty now have the word "all" to indicate that all 40 Principles should be tried (see Appendix).

To work with the matrix, follow these steps:

1. Write, or refer to, the contradiction statement in the form of IF …, THEN …, BUT …
2. Select the standard feature from Table 11.3 that best represents the feature that is improving within the initial contradiction. (THEN statement)
3. Select the standard feature from Table 11.3 nearest to the feature that will worsen in this case. (BUT statement)
4. Find the row in the matrix (Appendix) with the number of the improving standard feature.
5. Find the column in the matrix (Appendix) with the number of the worsening standard feature.
6. In the cell at the intersection of the row and column are the numbers of the principles that have been most frequently found useful to resolve this contradiction.
7. Look up those principles in the list of 40 Principles and use them to create ideas for solving your problem. Do not be surprised if some of the recommended principles do not help or if you find great ideas from principles that were not listed—the numbers in the cells are based on historical data, not on any knowledge of your particular problem. The listed principles are not best practices, and they may not be creative; they are the principles that other people have used most often.

Table 11.2 Selected Rows and Columns from the Contradiction Matrix

					Worsened Feature				
	Improved Feature	31	32	33	34	35	36	37	38	39
11	Stress or pressure	2, 33, 27, 18	1, 35, 16	11	2	35	19, 1, 35	2, 36, 37	35, 24	10, 14, 35, 37
12	Shape	35, 1	1, 32, 17, 28	32, 15, 26	2, 13, 1	1, 15, 29	16, 29, 1, 28	15, 13, 39	15, 1, 32	17, 26, 34, 10
13	Stability of object's composition	35, 40, 27, 39	35, 19	32, 35, 30	2, 35, 10, 16	35, 30, 34, 2	2, 35, 22, 26	35, 22, 39, 23	1, 8, 35	23, 35, 40, 3
14	Strength	15, 35, 22, 2	11, 3, 10, 32	32, 40, 25, 2	27, 11, 3	15, 3, 32	2, 13, 25, 28	27, 3, 15, 40	15	29, 35, 10, 14
15	Duration of action by a moving object	21, 39, 16, 22	27, 1, 4	12, 27	29, 10, 27	1, 35, 13	10, 4, 29, 15	19, 29, 39, 35	6, 10	35, 17, 14, 19

Note: The numbers in the cell refer to the principles that have the highest probability (based on historical analysis) of resolving the contradiction. For example, consider the proposal to change the speed of inflation of an airbag to reduce injuries to small occupants. The tradeoff is that injuries in high-speed accidents increase. Translating this into the TRIZ matrix terms, the parameter that improves is "Duration of action by a moving object" (Row 15) and the parameter that worsens is "Object-generated harmful effects" (Column 31). The cell at the intersection has the notation "21, 39, 16, 22." These are the identifiers for four of the principles of invention.

Table 11.3   Explanation of the 39 Features of the Contradiction Matrix[8]

	Title	*Explanation*
	Moving objects	Objects that can easily change position in space, either on their own or as a result of external forces. Vehicles and objects designed to be portable are the basic members of this class.
	Stationary objects	Objects that do not change position in space, either on their own or as a result of external forces. Consider the conditions under which the object is being used.
1	Weight of moving object	Mass of the object in a gravitational field. Force that the body exerts on its support or suspension.
2	Weight of stationary object	Mass of the object in a gravitational field. Force the body exerts on its support or suspension or on the surface on which it rests.
3	Length of moving object	Any one linear dimension, not necessarily the longest, is considered a length.
4	Length of stationary object	Same.
5	Area of moving object	A geometrical characteristic described by the part of a plane enclosed by a line. Part of a surface occupied by the object, or the square measure of the surface, either internal or external, of an object.
6	Area of stationary object	Same.
7	Volume of moving object	Cubic measure of space occupied by the object. Length × width × height for a rectangular object, height × area for a cylinder, etc.
8	Volume of stationary object	Same.
9	Speed	Velocity of an object. Rate of a process or action in time.
10	Force	Force measures the interaction between systems. In Newtonian physics, force = mass × acceleration. In TRIZ, force is any interaction that is intended to change an object's condition.
11	Stress or pressure	Force per unit area. Tension.
12	Shape	External contours, appearance of a system.
13	Stability of object's composition	Wholeness or integrity of the system. Relationship of system's constituent elements.Wear, chemical decomposition, and disassembly are all decreases in stability. Increasing entropy is decreasing stability.

(Continued)

Table 11.3 (Continued)   Explanation of the 39 Features of the Contradiction Matrix

14	Strength	Extent to which the object is able to resist changing in response to force. Resistance to breaking.
15	Duration of action by a moving object	Time that the object can perform the action. Service life. Mean time between failure is a measure of the duration of action. Durability
16	Duration of action by stationary object	Same.
17	Temperature	Thermal condition of the object or system. Loosely includes other thermal parameters, such as heat capacity, that affect the rate of change of temperature.
18	Illumination intensity	Light flux per unit area, also any other illumination characteristics of the system such as brightness, light quality, etc.
19	Use of energy by moving object	Measure of the object's capacity for doing work. In classical mechanics, energy is the product of force × distance. This includes the use of energy provided by the supersystem (such as electrical energy or heat.) Energy required to do a particular job.
20	Use of energy by stationary object	Same.
21	Power	Time rate at which work is performed. Rate of use of energy.
22	Loss of energy	Use of energy that does not contribute to the job being done (see 19). Note: Reducing the loss of energy sometimes requires techniques that differ from improving the use of energy, which is why this is a separate category.
23	Loss of substance	Partial or complete, permanent or temporary, loss of some of a system's materials, substances, parts, or subsystems.
24	Loss of information	Partial or complete, permanent or temporary, loss of data or access to data in or by a system. Frequently includes sensory data such as aroma, texture, etc.
25	Loss of time	Time is the duration of an activity. Improving the loss of time means reducing the time taken for the activity. "Cycle time reduction" is a common term.

(Continued)

Table 11.3 (Continued)   Explanation of the 39 Features of the Contradiction Matrix

26	Quantity of substance/ quantity of matter	The number or amount of a system's materials, substances, parts, or subsystems that might be changed fully or partially, permanently, or temporarily.
27	Reliability	A system's ability to perform its intended functions in predictable ways and conditions.
28	Measurement accuracy	Closeness of the measured value to the actual value of a property of a system. Reducing the error in a measurement increases the accuracy of the measurement.
29	Manufacturing precision	Extent to which the actual characteristics of the system or object match the specified or required characteristics.
30	External harm affects the object	Susceptibility of a system to externally generated (harmful) effects.
31	Object-generated harmful factors	A harmful effect reduces the efficiency or quality of the functioning of the object or system, generated by the object or system as part of its operation.
32	Ease of manufacture	Degree of facility, comfort, or effortlessness in manufacturing or fabricating object or system.
33	Ease of operation	Simplicity: The process is not easy if it requires many people, requires many steps in the operation, needs special tools, etc. "Hard" processes = low yield; "easy" processes = high yield; they are easy to do right.
34	Ease of repair	Quality characteristics such as convenience, comfort, simplicity, and time to repair faults, failures, or defects in a system.
35	Adaptability or versatility	The extent to which a system or object responds positively to external changes. A system that can be used in multiple ways in a variety of circumstances.
36	Device complexity	Number and diversity of elements and element interrelationships within a system. User may be an element of the system that increases the complexity. The difficulty of mastering the system is a measure of its complexity.
37	Difficulty of detecting and measuring	Measuring or monitoring systems that are complex and costly, require much time and labor to set up and use, or have complex relationships between components or components that interfere with each other all demonstrate "difficulty of detecting and measuring." Increasing cost of measuring to a satisfactory error is also a sign of increased difficulty of measuring.

(Continued)

Table 11.3 (Continued)   Explanation of the 39 Features of the Contradiction Matrix

| 38 | Extent of automation | The extent to which a system or object performs its functions without human interface. The lowest level of automation is the use of a manually operated tool. For intermediate levels, humans program the tool, observe its operation, and interrupt or reprogram as needed. For the highest level, the machine senses the operation needed, programs itself, and monitors its own operations. |
| 39 | Productivity | The number of functions or operations performed by a system per unit time. The time for a unit function or operation.The output per unit time or the cost per unit output. |

11.16.1.1 Principles Selection for Chapter 7 Case Studies

In Chapter 7, the 40 Principles selected for use in resolving the case studies (blacksmith, wireless power, and call center) were simply presented for use. The process by which the principles were selected was not discussed as the use of the matrix was not yet presented. Now that you know how to utilize the matrix to identify principles for use in tradeoff contradiction resolution, the selection pathways for the principles used to resolve the Chapter 7 case studies are shown:

11.16.1.1.1 Blacksmith (see Section 7.6.1)

Note—Only the first contradiction in this case study is used to demonstrate the selection of the 40 Principles. After studying the process here, see if you can use the methodology to uncover how the second and third contradictions were solved by way of selecting specific 40 Principles from the matrix.

1. Write, or refer to, the contradiction statement in the form of IF ..., THEN ..., BUT ...
   Tradeoff contradiction:
   ■ IF a heated metal piece is worked in an open area
   ■ THEN the work space is somewhat simplistic
   ■ BUT the ambient air will cool the piece too rapidly
2. Select the standard feature from Table 11.3 that best represents the feature that is improving within the initial contradiction.
   (THEN statement)

   "Then the work space is somewhat simplistic"

   The matrix feature chosen to present the THEN statement was "device complexity." Device complexity was chosen as the improving feature since the contradiction states " ... work space is somewhat simplistic."
3. Select the standard feature from Table 11.3 nearest to the feature that will worsen in this case.
   (BUT statement)

   "But the ambient air will cool the piece too rapidly"

The matrix feature chosen to present the BUT statement was "temperature." Temperature was chosen as the worsening feature since the concern is with the air cooling the work piece too rapidly.

(Note—Those two Contradiction Matrix features are not the only options. For example, another improving feature choice could be "ease of operation" while other worsening feature choices could be "loss of energy" or "object-generated harmful factor.")

- IF a heated metal piece is worked in an open area
- THEN the work space is somewhat simple: "device complexity"—36 (improving feature)
- BUT the ambient air will cool the piece too rapidly: "temperature"—17 (worsening feature)

4. Find the row in the matrix (Appendix) with the number (36) of the improving standard feature.
5. Find the column in the matrix (Appendix) with the number (17) of the worsening standard feature.
6. In the cell at the intersection of the row and column are the numbers of the principles that have been most frequently found useful to resolve this contradiction.

   Looking up "device complexity" on the matrix's Y-axis (improving feature) and "temperature" on the matrix's X-axis (worsening feature) resulting in the identification of principles:

   - Principle 2—Separation (taking out) (no solution direction triggered for the author by this principle—what ideas does it give you?)
   - Principle 17—Dimensionality Change (another dimension)
   - Principle 13—"The Other Way Around"

7. Look up those principles in the list of 40 Principles and use them to create ideas for solving your problem. See Section 7.6.1 for solution concepts applied to this case study by way of the identified 40 Principles.

11.16.1.1.2 Wireless power system improvement (see Section 7.7.1)

1. Write, or refer to, the contradiction statement in the form of IF ..., THEN ..., BUT ... Tradeoff contradiction:
   - IF small Tx and Rx coils are used in the wireless power system
   - THEN the coils will fit in the side panels of the computer and mobile phone
   - BUT the transmission distance of the wireless power will be very small
2. Select the standard feature from Table 11.3 that best represents the feature that is improving within the initial contradiction.
   (THEN statement)

   "Then the coils will fit in the side panels of the computer and mobile phone"

   The matrix feature chosen to present the THEN statement was "area of stationary object." Area was chosen as the transmitting and receiving antennas have little depth and can be considered areas. Stationary object was chosen because the antennas are stationary in relation to the devices (computer and mobile phones) in which they are installed.

3. Select the standard feature from Table 11.3 nearest to the feature that will worsen in this case.

(BUT statement)

"But the transmission distance of the wireless power will be very small"

The matrix feature chosen to present the BUT statement was "power." Power was chosen as it is the effectiveness of the power usage in charging the mobile phone that is of interest and when the antenna sizes are limited by the side panels of the computer and mobile phone, then the effectiveness is poor.

- IF small Tx and Rx coils are used in the wireless power system
- THEN the coils will fit in the side panels of the computer and mobile phone: "area of stationary object"—6 (improving parameter)
- BUT the transmission distance of the wireless power will be very small: power—21 (worsening parameter)

4. Find the row in the matrix (Appendix) with the number (6) of the improving standard feature.
5. Find the column in the matrix (Appendix) with the number (21) of the worsening standard feature.
6. In the cell at the intersection of the row and column are the numbers of the principles that have been most frequently found useful to resolve this contradiction.
Looking up "area of stationary object" on the matrix's Y-axis (improving feature) and "power" on the matrix's X-axis (worsening feature) resulting in the identification of principles:
   - Principle 17—Dimensionality Change (another dimension)
   - Principle 32—Optical Property Changes (changing the color)
7. Look up those principles in the list of 40 Principles and use them to create ideas for solving your problem. See Section 7.7.1 for solution concepts applied to this case study by way of the identified 40 Principles.

11.16.1.1.2 Call center improvement (see Section 7.7.4)

1. Write, or refer to, the contradiction statement in the form of IF …, THEN …, BUT … Tradeoff contradiction:
   - IF the supervisor creates the customer support procedures
   - THEN there is efficiency in procedure creation for all of the agents
   - BUT the supervisor may not be familiar enough with all of the potential customer issues to create comprehensive support procedures
2. Select the standard feature from Table 11.3 that best represents the feature that is improving within the initial contradiction.
(THEN statement)

"Then there is efficiency in procedure creation for all of the agents"

The matrix feature chosen to present the THEN statement was "loss of time." Loss of time was chosen since one of the benefits to having the supervisor write the procedures for all of the agents is that a single person writes the procedures (not each agent) which greatly improves the efficient use of time (the "loss of time" is reduced).

3. Select the standard feature from Table 11.3 nearest to the feature that will worsen in this case.
(BUT statement)

"But the supervisor may not be familiar enough with all of the potential customer issues to create comprehensive support procedures"

The matrix feature chosen to present the BUT statement was "loss of information." Loss of information was chosen since the negative effect of having the supervisor write all of the procedures is that the supervisor's knowledge of customers' issues would be understandably limited, and therefore the chances that the supervisor could write a comprehensive set of procedures is low.

- IF the supervisor creates the customer support procedures
- THEN there is efficiency in procedure creation for all of the agents: "loss of time"—25 (improving parameter)
- BUT the supervisor may not be familiar enough with all of the potential customer issues to create comprehensive support procedures: "loss of information"—24 (worsening parameter)

4. Find the row in the matrix (Appendix) with the number (25) of the improving standard feature.
5. Find the column in the matrix (Appendix) with the number (24) of the worsening standard feature.
6. In the cell at the intersection of the row and column are the numbers of the principles that have been most frequently found useful to resolve this contradiction.

Looking up "loss of time" on the matrix's Y-axis (improving feature) and "loss of information" on the matrix's X-axis (worsening feature) resulting in the identification of principles:

- Principle 24—Intermediary
- Principle 26—Copying
- Principle 28—Mechanical Interaction Substitution (no solution direction triggered for the author by this principle—what ideas does it give you?)
- Principle 32—Optical Property Changes (changing the color) (no solution direction triggered for the author by this principle—what ideas does it give you?)

7. Look up those principles in the list of 40 Principles and use them to create ideas for solving your problem. See Section 7.7.4 for solution concepts applied to this case study by way of the identified 40 Principles.

11.16.2 Inherent Contradictions and Resources

Principles can be suggested by inherent contradictions and by resources (considered in Chapters 4 and 5). The constraints of the problem may require that opposite requirements are satisfied at different times, in different places, or at the same time at the same place. The three upper rows in Table 11.4 present these situations.

Let us try these recommendations on some of the examples we have been working with.

- Big muffler–no muffler: Because the muffler should be big and have zero size at the same time in the same place, multifunctionality principle (6) is the first to think of.
- Many seeds–single seed: Because there should be many seeds and individual seeds at the same time in the same place, the merging principle (5) may be the first principle to try.

Table 11.4   Some Principles Related to Certain Contradictions and Resources

Contradiction and Resource	Principles
Principles for realizing incompatible requirements in different times (time resources)	Segmentation (1), preliminary counteraction (9), preliminary action (10), beforehand compensation (11), dynamic parts (15), periodic action (19), hurrying (21), discarding and recovering (34).
Principles for realizing incompatible requirements in different places (space resources)	Separation (2), local quality (3), symmetry change (4), nested doll (7), dimensionality change (17), flexible shells and thin films (30).
Principles for realizing incompatible requirements at the same time in the same place (resources on the macro- and microlevel)	Segmentation (1), merging (5), multifunctionality (6), intermediary (24), composite materials (40).
General resources: air, water, space, gravity, others	Pneumatics and hydraulics (29), weight compensation (8), inert atmosphere (39).

■ Big clearance–no clearance: Because the clearance should be big and zero at different times, segmentation (1) and dynamic parts (15) can be selected. Because the pin and the hole have simple geometric forms, geometric principles are also interesting.

■ Long training time–no training time: Training should be long and short at the same time in the same place. Segmentation (1) and multifunctionality (6) are good principles.

■ Much water–no water: There should be a lot of water to suppress the fire, and there should be no water to keep the equipment simple and to decrease water damage. Browsing the principles, we find porous materials (31). Porous suggests a lot of substance in the sense of volume, and little substance in sense of mass. "Porous water" could mean air bubbles in the water or water droplets in air. Very small droplets could be a mist filling the whole room. Therefore, there would be a lot of substance (big volume of mist) and, at the same time, very little substance (small mass of water). This also can be expressed as "segmentation" (1).

Depending on the system, a variety of general resources such as air, water, space, and gravity are available. The last row of Table 11.4 shows some principles that can help you to use these resources.

Always consider the components of the resource and modifications of the resource. If you have air, you have nitrogen, oxygen, carbon dioxide, and small concentrations of other gases. If you have sunlight, you have yellow light, red light, ultraviolet light, infrared light, and so on. If you have direct current electric power, you can make a loop in the wire carrying the electricity and make a magnet. If you have water and power, you could make steam or ice. Then, apply the principles to these new resources.

11.16.3 Using the Attributes of the Ideal Final Result

Principles can be selected by the attributes of the Ideal Final Result (considered in Chapter 6).

The ideal system has no weight, no size, and consumes no energy. Principles that have a logical relationship to the support of the increase of ideality are shown on the last row of Table 11.5. Some other component should deliver the function of the muffler or the pin. Some other activity, for example work, should include training. Seeding should be combined with other operations. Training should become self-training, and so on.

It will be particularly helpful to formulate the Ideal Final Result and why the ideal cannot be achieved. This will usually lead to contradictions. For example,[9] in a customer-service situation, the Ideal Final Result is that customers get exactly what they need exactly when they need it. An analysis of the problem might follow this path:

"I can't give it to her because my employees don't have all the knowledge."
"Why not?"
"Because employee turnover is so fast that untrained employees are used."

This analysis reveals several potential problems and families of solutions:

- Customers get what they need without the (direct) help of employees.
- Employees have the knowledge without training.
- Trained employees do not leave the job.

Table 11.5 Patterns of Evolution and Related Principles

Pattern	Principles
Uneven evolution of the system	All.
Transition to macrolevel	Merging (5), nested doll (7), weight compensation (8), multifunctionality (6), composite materials (40).
Transition to microlevel	Segmentation (1), nested doll (7), mechanical interaction substitution (28), flexible shells and thin films (30), porous materials (31), parameter changes (35), phase transition (36).
The increase of interactions: Introduction of substances and actions	Porous materials (31), inert atmosphere (39), feedback (23), intermediary (24), mechanical interaction substitution (28), thermal expansion (37), parameter changes (35).
Expansion and convolution	Separation (2), local quality (3), merging (5), multifunctionality (6).
Summary: Increasing of ideality	Multifunctionality (6), "Blessing in disguise" (22), self-service (25).

Table 11.6   Apply Innovative Principles to Your Own Problem

1. What is the contradiction?
2. If this contradiction, as expressed by the reader, can be reflected in the Contradiction Matrix, which principles are suggested?
3. If this is an inherent contradiction, which principles might help?
4. What ideas did you get from each principle?
5. Browse through all the principles. What additional ideas did you get?

Now use the 40 Principles to look for solutions to each of these categories of problems, then select one or more that you believe may help improve the situation. When applying the 40 Inventive Principles, keep in mind the TRIZ concepts of removing the reason for the contradiction and using available resources.

Principles can be selected by the patterns of evolution (considered in Chapter 10). Some principles are the same as the patterns and some others can be easily seen as ways to use patterns. Conversely, the patterns can be seen as arising from the repeated use of certain sets of principles. Table 11.5 illustrates the correspondence between patterns and principles: only the most obvious links are shown in the table. The more you use the principles, the more you will see connections between them and evolutionary patterns.

You might make different choices of principles to apply to the example problems. The examples shown here are not at all exhaustive. The principles are good ideas, based on many strong solutions to a wide variety of problems over many years. Which solutions will be best for your particular problem will depend on the problem, the resources, and the constraints. Try using the 40 Principles to get new insights that can help you find new ways to approach solving your problem (Table 11.6).

11.17 Summary

Innovative principles give clues for solving problems. The simplest way to use the principles is to go through them all or just randomly select them.

Some selection criteria are useful for deciding which principles to use for any particular problem. Seek principles that resolve tradeoffs. Try the Contradiction Matrix if it fits your tradeoffs.[10] Seek principles that eliminate the inherent contradictions. Resource mapping, the description of the Ideal Final Result, and the patterns of evolution also help to find useful principles.

There are a wide variety of 40 Principles listings written with examples associated with different disciplines, industries, and applications. A selection of these can be downloaded from the web page innomationcorp.com/simplifiedTRIZ. Other literature can provide additional insight to the 40 Principles and their application, for example Ian Seed's book.[11]

Acknowledgments

The authors would like to thank Darrell Mann and John Terninko for generously sharing their lists of examples of many of the 40 Principles. We also thank Joe Miller and Ellen MacGran for their help with the research for features of the matrix.

References

1. Altshuller, G. S. 1997. *40 Principles*. Worcester, MA: TIC.
2. Savransky, S. D. 2000. *Engineering of Creativity*. Boca Raton, FL: CRC Press, 199.
3. Zlotin, B. and Zusman, A. 1999. Managing innovation knowledge, *Izobretenie*, I, No. 21.
4. Altshuller, G. S. 1997. *40 Principles*. Worcester, MA: TIC, 125.
5. Mann, D. 2001. Using TRIZ to overcome the mass customization contradiction. Presented at TRIZ conference "TRIZ Future 2001," Bath, November 7–9, 2001.
6. Mann, D. and Domb, E. 1999. Business contradictions: 1) Mass customization. *The TRIZ Journal*, December 4. https://triz-journal.com/business-contradictions-1-mass-customization/.
7. Wright, M. and Patel, M., Eds. 2000. *How Things Work Today*. London: Marshall Publishing, 303.
8. Domb, E. 1998. The 39 features of Altshuller's Contradiction Matrix. *The Triz Journal*, November 21. https://triz-journal.com/39-features-altshullers-contradiction-matrix/.
9. Mann, D. and Domb, E. 1999. 40 inventive principles for business. *The Triz Journal*, September 1. https://triz-journal.com/40-inventive-business-principles-examples/.
10. Contradiction, Matrix. (n.d.). In Wikipedia. Retrieved June 22, 2017, from http://www.triz40.com/TRIZ_GB.php.
11. Seed, I. 2016. *Successful Problem Solving*. Reading, UK: Cogentus Consulting Limited, 103–236.

Chapter 12

Moving from Challenging Showstoppers to Innovative Solutions

12.1 Introduction

In Chapter 7, we introduced a few examples showing how functional analysis supported by TRIZ produced new solutions. In this chapter, we will share more solutions that have been generated by TRIZ. By solution, we mean innovation or the result that has been realized commercially.

12.2 Big Innovations in Paper Machinery

It is as in a Hollywood story, with difficulties before the happy end.

Prolific Finnish inventor Pekka Koivukunnas can share a lot about the implementation of inventions made by the support of TRIZ.

In the beginning of the 1990s, Koivukunnas worked in Valmet, the company producing paper machinery. Along with Juha Lipponen, Koivukunnas created an important invention that allowed an improvement to a traditional calender. Lipponen was then a graduate student preparing his M.Sc. thesis on calender technology. The topic was to design a new supercalender.

A supercalender is a machine that smooths paper at the end of a paper-manufacturing process. More specifically, it is a set of rotating rollers stacked on top of each other. It is big, the size of a small multistory building. The system contains two kinds of rollers. One roller is made of metal, the other of a metal core with a soft outer layer. The paper is moved between many pairs of the metal and soft rollers. The pressure between rollers makes the paper smooth. In principle, the calender works as a mangle (a machine for ironing laundry by passing it between heated rollers), only it smooths paper instead of clothing or cloth.

The traditional supercalender produced the needed pressure simply by the weight of the roller sets pressing down on each other as the paper went on a zigzag trajectory between the rollers. To improve quality, it was desirable to increase pressure between the rollers. However, the traditional wisdom of papermaking said that this solution would not work as the lowest rollers would be damaged if an additional load were added to the natural weight of the roller sets. In practice, engineers chose a compromise: the natural weight of rollers plus a small additional load. The result was unsatisfactory as there was too little pressure between the rollers higher in the stack.

Koivukunnas and Lipponen looked at the problem from another point of view. Koivukunnas, who knew the methods of TRIZ, asked Lipponen to formulate the problem as a contradiction. The question was, "How to make even pressure between all the roller sets throughout the stack?" Lipponen thought of the problem and proposed support-bearing housings so that there was even pressure between all of the rollers. Why not regulate pressure by forces that can be easily created through load-bearing housings?

Then, the real contradiction began to appear. Support by bearing housings was impossible because of different deformations of the hard and soft rollers. The system would create inferior quality since the rollers would pressure the paper unevenly. To successfully implement this solution concept, the rollers should have absolute stiffness along their entire length. But this was not possible as the long and heavy rollers, supported only by bearing housings at both ends, would inevitably be deformed by gravity.

The contradictory requirements were finally understood—the rollers should be stiff and inflexible, to get even pressure, and they must be flexible, due to the laws of physics.

The question was, "How to get even pressure although the long rollers would become deformed?"

In July 1993, Koivukunnas proposed one possibility in resolving the contradiction. What would happen if all rollers "hung" uniformly? The rollers would then receive even contact across their entire length. And the load would be even throughout the whole stack of the rollers. This would be possible if every free-hanging roller had the same deformation. Constant deformation was the core of the invention. The solution of the contradiction, or the Ideal Final Result, was that the flexible rollers would behave as if they were stiff. One can also say that rollers are flexible as a whole and yet stiff at every point.

The "impossible" contradiction was solved by simple and elegant mechanics. A good solution seems simple and uses the laws of nature even if, at first, it seems that the laws of nature will prevent a good solution.

Koivukunnas found that the uneven deformation, and therefore uneven pressure, could be avoided by using identical rollers instead of the alternating hard and soft rollers. As such, he remembered that there already existed a so-called soft calender for the production of office papers. A variation of the soft calender, a roll of thin soft layer on metal core, could be used on both sides of the paper web. The layer of soft material is effectively hard when under pressure.

Experiments showed that the new solution worked as was patented. Even though an effective solution existed, it was not initially accepted in the marketplace. Sales people considered it useless since the process run with old calenders seemed to perform acceptably.

Then, one day, the largest competitor of Valmet unveiled a similar solution. Valmet's top management then required a quick implementation of the new design.

Clearly, the competitor had read Valmet's patent, but could not use the protected idea of uniform deformation or rollers. Only an additional load was used. The set of rollers were designed to carry the sum of gravity and the extra weight. However, problems arose as rollers were broken. The weaknesses of the compromised solution became evident. Valmet's invention won the silent competition, and the invention was given the brand name of OptiLoad, conquering the market.

The sales people were now also happy as the economic impact of OptiLoad was in the order of one billion US dollars over the time of the patent protection.

The innovator, however, thought further: "The sales staff said that customers wanted more pressure, temperature, speed and the width of the web. But, more pressure caused the negative effect of paper weakness." Koivukunnas said, "I once again thought of the problem through the TRIZ framework including contradictions and resources."

The contradiction: High pressure produces a smooth surface (good), but the paper output is too thin (bad). Low pressure produces the proper thickness of paper (good), but the paper output has a rough surface (bad).

How could these contradictions be resolved? How could the pressure be both high and low simultaneously? Generally, how can something or some action be both high/big and low/small simultaneously?

The abstract answer: A big thing or action can be comprised of many small ones.

Ideal Final Result: The rollers produce the required thickness and smoothness of paper with minimal pressure.

The solution concept came from TRIZ Principle 13—"The Other Way Around": less pressure, not more. This idea was, of course, contrary to methods taken for granted for decades in the paper industry.

Resources: It happened that a solution exactly along these lines was already arrived at by a colleague, Mika Viljanmaa: one roller followed by a metal belt instead of additional rollers. The belt, having a much larger surface area than a roller, allows the paper to be smoothed with low pressure.

It is important to point out that TRIZ can also lead us to existing inventions that suffice our requirements, as demonstrated here. This is one benefit of the methodology.

In the beginning, the management once again rejected this new idea. Therefore, the process engineers made their first experiments in secret.

After the engineers' successful experiments, management changed their policy. Metal belt calendering is now used successfully and the device, OptiCalender Metal Belt, is on the market. In 2012, the king of Sweden presented Mika Viljanmaa with the Marcus Wallenberg prize, considered by many the "Nobel Prize" in the forest products industry. The Wallenberg Foundation announced that the metal belt technology can give global annual savings in the order of 2–3 billion USD.

Some lessons learned from this case study:

1. Problems need to be well formulated in order to get good solutions. Original problem statements are often insufficient or even wrong.
2. Good ideas often direct us to existing solutions that can be adapted to your requirements.
3. Good ideas, fulfilling the requirements of the Ideal Final Result, are robust and will eventually be utilized, even if initially meeting resistance. The resistance is not a good or bad thing; it is simply a fact. Often, we will see the following progression:
   a. You have a good idea. They say, "It doesn't work."
   b. You show that it works. They say, "It is too expensive."
   c. You show that it works and is cost-effective. They say, "No one needs it."
   d. You show that it works, is profitable, and people love it. They say, "We thought of it first."
4. Obtaining management buy in before solving a problem will usually avoid organizational implementation issues later on.

12.3 Small Innovations in Semiconductor Process Equipment: 50 USD Solution Saves Millions of Dollars

In 2009, the author was leading a beginner TRIZ training class at Intel Corporation, where multiple teams brought their own problems to be solved. As to be discussed in Chapter 15—How to Drive the Adoption of TRIZ in Your Organization, Intel was utilizing the method of expert TRIZ practitioners leading novice practitioners through the application of TRIZ tools to challenges the novices were facing in their specific operational areas. One such team was working on an issue involving maintenance procedures for a piece of processing equipment called a "rapid thermal annealer" (RTA). The RTA utilized a large heating chamber where temperature probes, along with other sensing equipment, were inserted through the walls of the chamber (see Figure 12.1—Operational configuration). The temperature probes utilized sapphire-tipped temperature sensors. The small diameter (approximately one-eighth of an inch) cylindrical sapphire tips protruded from their associated temperature probe without any shielding or protection so that only the sensor (sapphire tip) extended into the chamber, and past the reflector shield, to ensure that chamber environment was minimally disrupted and the sensors could take accurate readings.

Part of the maintenance procedure was to remove each temperature sensor (there were 10 or more on each RTA chamber) and check them for damage, and then clean, calibrate, and reinsert them into the RTA heating chambers. The problem was that the sapphire tips were very brittle and would often crack, chip, or break if they even slightly grazed the chamber penetrations walls as they were removed and reinserted into the round (approximately three-sixteenths of an inch) chamber penetration holes (see Figure 12.1—Damage mode). Not until the RTA requalification and testing process (required to recertify the equipment for manufacturing after the completion of the maintenance procedures) was it possible to tell if a probe tip had even been damaged. The probe tips were so brittle, in fact, that on average half of the tips (over 350 a year) were damaged during the maintenance process. A damaged tip would produce readings of 25°C higher than actual. Further, each temperature probe cost 5600 US dollars, which totaled almost 2 million US dollars per year in additional maintenance expense across all of Intel's factories worldwide. The

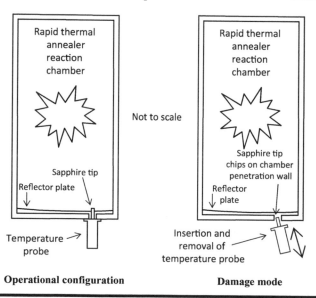

Figure 12.1   RTA chamber with sapphire-tipped temperature probe.

goal of the project was to substantially reduce, or more ideally eliminate, any damage to the sapphire tips during the maintenance process.

The project team produced a functional model of their system to focus in on the contradictory requirements of the maintenance procedure of removing and reinserting the temperature probes (see Figure 12.2). As discussed in Chapter 7—Understanding How Systems Work: Utilizing Functional Analysis to Expand Knowledge About Your Problem, the trouble areas (zones of conflict) are where harmful, excessive, or insufficient functions occur between components. As can be seen from the functional model, it was the relationship between the sapphire tip and the chamber/reflector plate where most of the trouble occurred. Therefore, a tradeoff contradiction statement was written about those harmful and insufficient relationships:

- IF the temperature probe is inserted into the chamber as an entire unit
- THEN the temperature probe does not have to be disassembled for the procedure
  - Associated improving feature: "Stability of object's composition"
- BUT the sapphire tip can be easily damaged during the procedure
  - Associated worsening feature: "Object-affected harmful factor" (also known as "external harm affects the object")

Principles identified from the Contradiction Matrix:

- Principle 35—Parameter Changes (no ideas generated from this principle)
- Principle 24—Intermediary
- Principle 30—Flexible Shells and Thin Films
- Principle 18—Mechanical Vibration (no ideas generated from this principle)

Utilizing Principles 24 and 30, the team developed an almost perfectly ideal solution. In fact, the solution was so elegant, simplistic, and cheap that the problem was 100% eliminated, and the total cost of implementation across all of Intel's nine sites worldwide was only 50 US dollars.

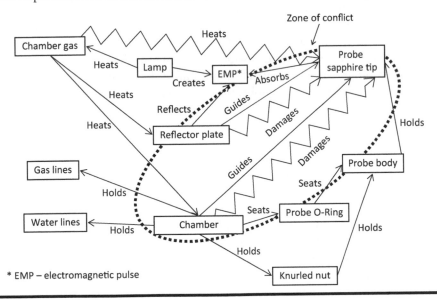

Figure 12.2   Functional model of RTA with temperature probe.

Utilizing Principles 24 and 30, see if you can also solve their problem before reading the elegantly simple solution they developed.

    Hint: The heating chamber could also be accessed from the inside during the maintenance procedure.

    Principle 24—Intermediary

Use an intermediary carrier article or intermediary process
Merge one object temporarily with another (which can be easily removed)

Ideas: _____

_____

_____

_____

Principle 30—Flexible Shells and Thin Films

Use flexible shells and thin films instead of three-dimensional structures
Isolate the object from the external environment using flexible shells and thin films

Ideas: _____

_____

_____

_____

The solution developed by the Intel project team was so brilliantly simple it still makes the author smile. Principle 24—Intermediary suggests to use something between the interfering components and be sure that it can be easily removed. Principle 30—Flexible Shells and Thin Films says that that intermediary object should be a flexible shell or thin film.

The solution was to find a piece of flexible hosing with an inner diameter just larger than the sapphire tip and an outer diameter just smaller than the chamber penetration hole (see Figure 12.3). During removal of the temperature probes, the hosing was inserted down through the chamber penetration and over the sapphire tip. Then, the temperature probe could be easily pulled out of the chamber penetration without risking chipping or cracking the sapphire tip. Next, the sapphire tip was pulled out of the hose (which had been pulled down through the penetration as the temperature probe was also unattached and pulled down and away from the chamber). After maintenance of the temperature probe, the sapphire tip was then reinserted into the end of the hose, and the hose was pulled back up through the chamber penetration hole (from within the chamber) as the temperature probe was lifted back into its attachment position. This method protected the sapphire tip from chipping or cracking by eliminating any contact, during removal or insertion, between the sapphire tip and the chamber penetration hole. After the temperature probe was reattached to the chamber, the hose was pulled off the sapphire tip leaving it undamaged and in position for normal operations.

    Note: It should be discussed that someone familiar with TRIZ might claim that the use of the matrix's improving feature of "stability of object's composition," by the project team, is not a very accurate assignment of that feature to this situation. Some would say that the parameter of "stability of object's composition" is meant for changes in stability of composition as a result of operational activities, not as a result of disassembly, and was, therefore, incorrectly applied, and thus the results of the analysis

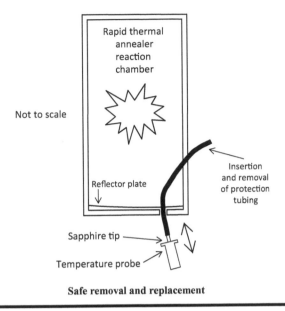

Figure 12.3   Sapphire tip protection solution.

were invalid. Many consultants in the TRIZ business focus on whether the analyses are executed according to their liking and less on the results of those analyses. There is a contrary viewpoint, "if the problem is solved and value is increased, then the analysis was not incorrect."

Some lessons learned from this case study:

1. Contradiction modeling is key in properly defining a problem.
2. First-time users of TRIZ can produce astonishing results (this TRIZ analysis was the very first one attempted by the maintenance technicians).
3. While the "solution" hose ultimately defined for use throughout all of Intel's factories worldwide was standardized, the initial hose used to prove the concept was a locally available resource as it was found in one of the technician's tool boxes.
4. The use of TRIZ methods is more flexible then some would lead you to believe: if you solve your problem with an increase in ideality, you did it right.

12.4 Summary

Proper problem definition is crucial to good solution generation. Several iterations of problem definition may be necessary to arrive at the best description of the problem.

Having management commitment for problem-solving projects is important during the implementation phases of the project. If solutions are not implemented, then solution generation has no value.

TRIZ beginners can produce astounding and ideal results. Understanding existing issues through the new lens of contradiction analysis often produces "obvious" solutions that are cheap, simple, and effective.

Chapter 13

TRIZ Knowledge Helps Drive Improvement

13.1 Introduction

In previous chapters, we have spoken of contradictions, functional modeling, resource analysis, ideality, and many other solution-guiding techniques. In short, there are many tool sets within TRIZ that help us to "draw" our solution vector in the correct direction. However, the more you use TRIZ, the more opportunities you will find to practice developing innovations by shooting "from the hip," and the easier you will be able to apply your new TRIZ knowledge. In other words, some TRIZ tools, especially patterns and principles, can be mentally accessed "on the fly" to imagine advancements for systems that you use daily.

This chapter will present some recent advancements and then utilize TRIZ tools to help the visualization of the next step of the advancement of those, and similar, systems. However, when studying this chapter, it is important to remember that the creation of solid innovations requires a bit of discipline and understanding of how the TRIZ tools work together (e.g., Is the contradiction I created focused on the root cause? Does the resolution of the contradiction utilize locally available resources and increase ideality?, etc.). With that in mind, it is alright to practice individual TRIZ tools, as outlined here.

13.2 Pay Attention to What's Around You

Many years ago, one of the authors was walking by a shop window and upon seeing a sewing machine had the thought, "there is an opportunity for a valuable merger of computing technology and traditional home appliances." An article discussing "the future modern sewing machine" was written for a local newspaper. This prompted the same thought around a similar transition in every kind of device.

Today, there are many home appliances with much more advanced information and control systems than their predecessors thanks to the pattern of transition to the macrosystems describing the merger of home appliances and microelectronics. What other "home appliances" might this concept be applied to, and what might the application look like? Study and fill in Table 13.1 to investigate other possibilities.

There are many cases throughout history that demonstrate an application of a TRIZ pattern or principle. Eric von Hippel and Georg von Krogh give an example of luggage handling.[1] In 1970, Bernard Sadow was lugging a heavy suitcase in an airport. While waiting at customs, he observed a worker rolling a heavy machine on a wheeled skid. He immediately had an idea. Why not add wheels to a suitcase? While the idea was not directly generated by way of a TRIZ tool, it does demonstrate the application of Principle 14—Curvature Increase (spheroidality). What other applications of Principle 14 can be used with luggage? Study and fill in Table 13.2 to investigate other possibilities.

There are many more examples:

■ Technologies that allow mobile phones to monitor breathing are being developed. They allow improved health care with an inexpensive cost. This example can be related to the pattern of increasing interaction. Study and fill in Table 13.3 to investigate other possibilities.
■ Plant-based food-packing materials, for example PLA (polyactic acid) and bio-PET (poly-ethylene terephthalate), decrease food losses with an ecological mindset. This example can be related to Principle 33—Homogeneity. Study and fill in Table 13.4 to investigate other possibilities.

Structural electronics (electronics embedded in load-bearing structures) are one of the important technological developments in our century. What pattern or principle does this represent and how can it be expanded further?

Pattern or principle_____

Additional application options_____

_____

Wind-turbine noise can be decreased by copying the structure of owl feathers and applying it to the turbine blades, a solution that illustrates the transfer of technologies across industries,

Table 13.1   Merger of "Home Appliances" and Microelectronics

System	Transition
Hammer (example)	Read out that reports striking force and angle
Faucet (example)	Input that allows choice of water temperature and readout that shows current temperature and flow rate
Window	
Vacuum Cleaner	
Others?	

Table 13.2   Application of Principle 14—Curvature Increase to Luggage

System	Transition
Luggage (example)	Put wheels on suitcase
Luggage wheels (example)	Replace wheels with ball roller
Luggage ball rollers (example)	Put ball rollers on swivels
Luggage handle (example)	Create adjustable curved ergonomic handle
Luggage structure (example)	Create curved ribbed structure that is rigid when luggage is opened and allows flexibility (stuff sack) when closed
Luggage shape	
Luggage size	
Others?	

Table 13.3   Application of Pattern of Increasing Interaction to Smart Phones (Medical)

System	Transition
Smart Phone Application (example)	Integrate breathing monitoring
Smart Phone Application (example)	Integrate eye sight testing
Smart Phone Application (example)	Integrate blood sugar level testing
Smart Phone Application?	
Smart Phone Connectivity (example)	Like a hard drive on-line back-up system, have smart phone regularly upload data to on-line medical records
Smart Phone Connectivity?	
Others?	

including technologies invented by nature (biomimetics). Altshuller was a big proponent of using the designs in nature to solve our problems. What other applications of biomimetics can you think of?

"Natural system" doner _____

Application ideas_____

Table 13.4   Application of Principle 33—Homogeneity to _____

System	Transition
Food packaging (example)	Create food packaging out of food based materials
Roadway and Tires (example)	Make tires out of asphalt or roadways out of synthetic rubber
Bandages (example)	Like "cat gut" sutures, make bandages out of natural materials which are ultimately absorbed by the skin
Garbage bags	
Window screens	
Contact lens	
Others?	

13.3 Summary

Some TRIZ tools and methods can be applied "on the fly." It is fairly straightforward to look at a system advancement (e.g., traffic signals that sense traffic flow and respond accordingly) and think which pattern or principle is at work. Then, you can try applying the same pattern or principle at the next level in the system to ideate new advances (e.g., traffic signals that directly control automated "self-driving" vehicles). Frankly, the same is true of many TRIZ tools. For example, it is easy to envision the ideal state of any system, think about unused resources in and around a system and, of course, as discussed in previous chapters, create a contradiction statement about a situation or problem. In other words, you do not have to sit down and spend substantial time performing TRIZ work. Try it the next time you are commuting to work. You will be amazed at how many "on the fly" improvement opportunities there are.

Reference

1. von Hippel, E. and von Krogh, G. 2016. CROSSROADS—Identifying viable "need–solution pairs": Problem solving without problem formulation. *Organization Science*, 27:3, 207–221. http://dx.doi.org/10.1287/orsc.2015.1023.

Chapter 14

Evaluation of the Model for Problem Solving

In the Preface, we stressed that the reader should test and refine the generic model for problem solving. Accept TRIZ because it works, not because it is in fashion.

When you use TRIZ, you will have two ways to evaluate the process:

1. Evaluate the results of implementation against the criteria of the Ideal Final Result, as studied in Chapter 8.
2. Evaluate the model and tools of TRIZ against your own accumulated knowledge and experience.

Using the model for problem solving provides a structured way to arrange and organize the TRIZ tools and gives you a system for creating new knowledge. Contradiction, ideality, and other basic concepts are open for further analysis, deepening, and implementation in new areas. The model brings a component of research to your work. TRIZ is evolving continuously; it is not a frozen dogma that does not change and is used without question. One should ask repeatedly: "Why this model? Why this tool? Why not something else?"

Each time that you use TRIZ to solve a problem, you have the opportunity to improve your own learning and to help improve TRIZ for others. Some people add yet another question to the agenda in Chapter 9: "What did I do well in this application of TRIZ? What should I do differently next time?" This self-analysis technique is also embedded in the ARIZ process (part 9 of ARIZ version 85C) mentioned in Chapter 9.

We hope that you have implemented the models and tools in this book to create innovative solutions to your problems. When you have done that, you can use the exercise in Table 14.1 to contribute to the development of TRIZ.

You can share these thoughts within your own organization and improve the version of TRIZ that you are using, or you can share them with the global TRIZ community. You are encouraged to send your suggestions to the authors at http://www.innomationcorp.com/simplifiedTRIZ.

Table 14.1 Exercise—Evaluation of the Model for Problem Solving

1.	Which concepts, models, and tools are most valid and useful for my work?
2.	Which points may require further development?
3.	Other thoughts.

Chapter 15

How to Drive the Adoption of TRIZ in Your Organization

The ever-growing successful implementation of TRIZ in different companies, organizations, countries, and cultures makes it easier to introduce to your organization. While saying this, we should remember that even in an ideal situation, there will be some resistance. We often hear of new tools and methodologies including many that may not work as advertised. It is important to keep this in consideration when anticipating the organizational obstacles, and how to overcome them, toward the introduction of new methods.

If you are a manager or a consultant and have heard of TRIZ, you probably have a question: "How can I prove that TRIZ really works and will improve our business?"

If you have already worked your way through the first 14 chapters and tried the exercises, you are already proficient at using TRIZ. This generates a new problem: how to help other people in the organization adopt TRIZ.

Those of you who work in organizations, whether they are private companies, government agencies, or mom-and-pop operations, should take TRIZ back to your coworkers to get the benefit of solving your own problems in your own environment. Private companies will be looking for a competitive advantage—to patent, to set new standards, or to get the proprietary advantage from being first to market with new concepts—and public organizations will be seeking maximum benefit for minimum cost. Both sets of goals need the breakthrough creativity of TRIZ.

This chapter, and those that follow, will help you in all these cases. It was discussed in Chapter 9 that the whole book can be divided roughly into three parts:

1. How to solve problems by identifying, analyzing, and resolving contradictions (Chapters 2 through 8).
2. The improvement of systems using the patterns of evolution and 40 Principles (Chapters 10 through 11).
3. The business applications of TRIZ (Chapters 12 through 17), where the systematic implementation process for TRIZ is presented to make the benefits of TRIZ quickly available to your organization.

In the last section, we discuss how to integrate TRIZ with other tools and methodologies. Chapter 16 demonstrates in detail the use of TRIZ with Six Sigma and other important quality-improvement methodologies.

In this chapter, we first consider the most typical obstacles to adoption of TRIZ. Second, we explain how to overcome the obstacles and present a flowchart for the introduction of TRIZ. Third, we study the three main steps of the introduction.

15.1 Typical Obstacles to the Adoption of TRIZ

The same obstacles to the implementation of TRIZ, and any new methodology for that matter, are met in large and small, public and private organizations. The ways to overcome the obstacles are very similar, which makes knowledge of these typical obstacles very useful.

TRIZ will give your organization the capability for breakthrough solutions to difficult problems. TRIZ radically enhances the quality and quantity of idea generation. Reading this statement, people have one of two reactions:

1. If it sounds too good to be true, it probably is. It is just one more piece of hype. It is the management training "flavor of the month."
2. Great! Let us get it in here and start everybody using it immediately.

If you have used TRIZ to generate creative solutions to your own problems, you will personally avoid reaction number one. But what can you offer to people who have not experienced the power of TRIZ?

The primary obstacles to organization-wide adoption of TRIZ are human, not technical:

- Time: People are too busy fighting fires to learn new methods of fire prevention.
- Suspicion: Other "new methods" have overpromised productivity improvement, customer satisfaction, faster time to market, higher return on investment (ROI), or economic value added (EVA), and such.
- Traditional systems of project management: If traditional milestones measure the success of a project, and the new process does not match those milestones, there will be great pressure to work within the existing system. For example, some product-development systems have a period dedicated to conducting tradeoff studies. TRIZ tells us that good solutions to problems avoid tradeoffs. If the organization continues to mandate the use of its traditional system, the use of TRIZ will be discouraged.
- NIH syndrome: As we know from Chapter 1, NIH means "not invented here" and can have double meaning for TRIZ—both "not invented in the organization" and "not invented in our country."
- "Well, it may work for so-and-so, but it won't work for us": "Our problems are different/high tech/not in their database/controlled by regulators, etc." This comment is a subcategory of NIH.

15.2 How to Introduce TRIZ into Your Organization

The flowchart in Figure 15.1 describes a structured method of introducing TRIZ into organizations that overcomes these obstacles. This method uses no "tricks" of cultural change or subtleties of organizational dynamics. It gets the professionals and managers in product design, service,

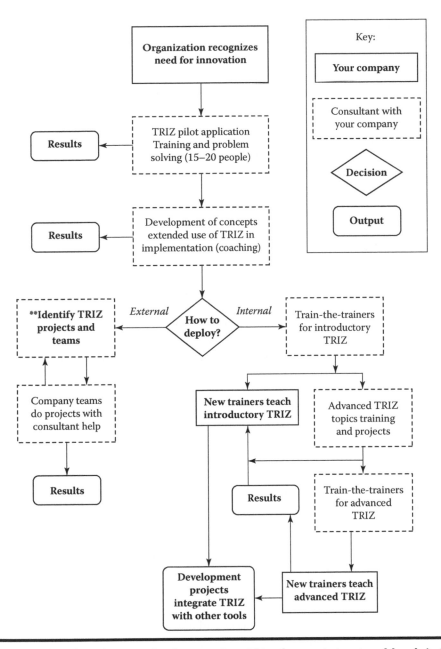

Figure 15.1   Flowchart for TRIZ implementation. This, the most structured level, is for large companies that want to become completely self-sufficient. Companies that want to use TRIZ quickly without studying it will start at the box marked **. The internal path is for the development of TRIZ experts inside the organization. The external path uses outside experts. Choosing whether to use consultants and outside experts is discussed in the text.

engineering, R&D, production, and distribution to experience TRIZ immediately and helps them get breakthrough results for their own problems quickly. Of course, some organizations are far less structured than this—we have even seen success in places where the company just buys everyone a copy of this book.

The effect of these immediate increases in creativity is that the obstacles labeled "suspicion" and "NIH" are removed, and the organization then uses its own resources (and its enhanced creativity) to reallocate the time of key people. In the early stages of TRIZ implementation, having a lot of new ideas is not always seen as a benefit because the organization may lack the resources to follow up on all of them.[1] The rationale for the process represented by the flowchart will be clear if each step is analyzed in terms of its direct results—new ideas, new concepts, creativity improvements—and the organizational change results.

15.3 Implementing the Steps of the Flowchart

The flowchart is a detailed outline of the steps that lead to successful TRIZ implementation. Three major steps encompass all the details:

1. The decision is made that increased innovation is needed.
2. Pilot projects: TRIZ and methods of teaching TRIZ are tested on the organization's projects, with its people working in their own surroundings. Implementation of the results of the pilot projects is a key to success because it will let the entire organization see how the TRIZ solutions work in practice.
3. Acceptance: TRIZ becomes part of the normal methods of operation for the organization.

15.3.1 Step 1

Step 1 is the organization's decision that increased innovation is needed. Most commonly, in private business, this comes because of competitive pressure or, in public agencies, from citizen demand, although occasionally it comes from regulatory requirements. One or more organization managers are selected to be a TRIZ "champion" to orchestrate the introduction and institutionalization of TRIZ. Frequently, these champions are people who have learned TRIZ and understand how it will help the organization and have used their knowledge of the organization's politics to get themselves appointed to do the job. It is more important that the champion be a respected person who understands how to get things done within the organization than to be highly knowledgeable in TRIZ.

It is essential for the champion to clarify needs and deal with the obstacles. One way to start is by asking and answering these questions:

■ Do we really need increased innovation? Champions may have to answer this question first, or get the leadership of the company to answer it, before dealing with TRIZ as the systematic way to increase innovation. The motivation for increased innovation could come from customers, from competitors, or from a regulatory situation. Champions will use the answer to this question to help people understand why a new method is being introduced into the organization.
■ Will TRIZ work in our circumstances? Companies that have been the first to implement generic models and templates of TRIZ in their industries (electronics, chemical,

medical, food, financial services, insurance, and others) have gotten great results. Because the TRIZ examples in existing textbooks did not deal with their particular industry, they had to spend some time understanding how to apply it to their situation. Now that TRIZ has been applied in so many different circumstances, this problem is somewhat lessened, but the champions may still have to find examples to convince others that this new method works. As an example, in late 2016 one of the authors supported a materials manufacturer in improving their production processes. The manufacturer used chemical inputs to produce product outputs for use by other manufacturers and industries. The work required an in-depth understanding of both engineering systems (production equipment) and the manufacturing process' input materials (chemicals). As such, it was necessary to initially model their systems as both equipment and chemical systems with a focus on functionally modeling the chemistry within those systems. Though the author has also executed chemistry equation functional modeling for the pharmaceutical industry, this simultaneous application of both chemistry and engineering problem analysis was a new application of functional modeling. Therefore, during the initial stages of program development for the company, pharmaceutical-based-type functional modeling was used to demonstrate application possibilities. In other words, analogous applications were used to help demonstrate that these new techniques could be moved into "new" industries.

In support of "selling" TRIZ to your organization, the following is a manifold listing of a few TRIZ applications that the authors have been directly involved in:

- Design of prototype hardware for quantum computing engines: Sandia National Labs
- Development of more robust satellite control systems: Air Force Research Lab
- Contradiction resolution of issues with satellite super structures: Air Force Research Lab
- Exploration of image-correction methodologies for curved glass displays: Samsung Electronics
- Improvement of microprocessor-testing methodologies: Intel
- Improvement of corn-breeding methods: A worldwide agribusiness
- Product line–issue elimination: Automotive A/C equipment supplier
- Improvement of utilization of operating room suites: Presbyterian Hospital
- Reduction, and in some instances, elimination, of drug side effects: International pharmaceutical company
- Analysis of the reduction of harm to terrestrial populations during humanitarian air drops: US Air Force
- Exploration of how to stop a speeding vehicle without damaging the vehicle or harming its occupants: US Air Force
- Training to focus software companies to better understand their full operational environment and, therefore, how to write coding that more efficiently runs that environment
- How to better distribute patients throughout a network of dispersed urgent-care clinics: Presbyterian Hospital
- How to improve in-network physician referral rate: Presbyterian Hospital

There are also many studies and reports that further document the successful implementation of TRIZ worldwide:

- Spreafico and Russo conducted a review of more than 200 industrial case studies executed in many countries between 1997 and 2014 where TRIZ was mostly used for quality improvement. The most utilized TRIZ tools were contradictions and the inventive principles.[2]
- Shaughnessy described in *Forbes* the successes of TRIZ within Samsung from the year 2000 on. Examples include DVD pick-up innovation and Super AMOLED displays.[3]
- Yeo *et al.* from Samsung Electronics have presented the study of nearly 200 projects: " ... analysis of 198 projects showed that the majority of those projects led to new patents and ideas."[4]
- China's National Institute of Clean-and-Low-Carbon Energy (NICE) is using TRIZ to increase efficiency and reduce environmental impacts in coal technology. "In 2014, NICE received a China Quality Management Benchmark Award because we successfully applied DFSS and TRIZ," says Yongwei Sun (TRIZ promoter in NICE.)[5]

Once you have received approval to begin implementation of a TRIZ program within your organization, you will find that commitment generates results and results strengthen commitment. The system presented here is designed to get results fast so the commitment can grow. The best tools will not give good results if there is no clear understanding of the need to change from the previous way of doing things to a new system. This list of questions may help clarify the need to change:

- What are the actual problems in our organization? (e.g., difficulties anticipating customers' needs, challenges in maintaining manufacturing quality.)
- What are our strengths? (e.g., good knowledge in science and engineering, strong financial analysis skills, excellent distribution management, reputation for a caring attitude, as well as strong medical skills)
- What are our weaknesses? (e.g., high-level experts have trouble finding simple solutions that customers understand; services and products are introduced with failure modes that cause customer problems; not all employees understand customer service is a priority.)
- What is needed to enhance strengths and remove weaknesses? (e.g., providing experts with models that help them use their knowledge more effectively.)

Also see Chapter 1—Why Do People Seek New Ways to Solve Problems?

15.3.2 *Step 2*

Step 2 is the selection of one or more pilot projects for TRIZ introduction. Candidate pilot projects can come from competitive situations, regulatory changes, or from the organization's problem identification and corrective-action system. The "champion" selects those problems that will have the best combination of high-value payoff and usefulness as future teaching cases. Good projects are those that are regarded as hard problems worth solving and that have management commitment to do so. These projects or problems are used as the case studies for an introductory class. Sometimes, the champion asks each class member to select a problem. The following memo is a template that the author has provided to many companies that can be modified and used to help people select good projects to bring to their first TRIZ class:

TRIZ Project Selection

TRIZ has many techniques for finding innovative solutions to hard problems in product, process, and transaction situations. Here is the list of characteristics of "good" problems for the TRIZ classes:

- We know who the customer is, what the customer's needs are, and why the present system is not satisfying those needs.
- We understand the root cause of the problem, not just the symptoms. BUT we don't know what to do.

    Why don't we know what to do?

    Sometimes, it is because of the presence of contradictions in the system.

    Sometimes, it is because the system has reached the limits of what can be achieved with the current technology or methodology—strength of materials, bandwidth, communications, or ...

    Sometimes, we don't know where to start.

    In your TRIZ class next week, you will not only learn the TRIZ methods, you will apply those methods to solve problems. Please bring anything you need to explain the problem—documents, drawings, etc. To give you some ideas about the kinds of problems that you could consider, here is a list of problems that have been solved by people in previous TRIZ classes:

    - Transaction: The project managers complain that the project management software takes too much time to use, so they don't keep the data current. Then, they lose the ability to do dynamic scheduling because their data is not current. Find a way to keep all the data current without taking extra time.
    - Product: A product has two customer adjustable settings that allow for a very wide array of product applications. The wide array of applications is good, but the combination of available settings is large enough that many customers become confused and only use the product in one mode. How can the next version of the product allow for a wider variation in usage without requiring the customer to master a somewhat complicated series of settings?
    - Transaction: The old telephone system for a large consulting company kept track of each call so that the appropriate client could be billed for the expenses. The new system doesn't have that capability, but some of the contracts require separate billing of phone calls. (Should have thought of this before installing the new system.) What can be done?

- Business: The company has a standard method for deciding which new projects should be funded. Many people think it is too complicated and they find ways to bypass it, causing great confusion about which projects are funded and about how the decisions are made. What can be done?
    - Measurement: The process has been through several cycles of improvement, and the yield has increased by several orders of magnitude. Measuring of defects is required by the customer, and it now takes a very large sample of the material to find enough defects to measure. Find a way to get a measurement that will be accepted by the customer without sacrificing a large quantity of product.

- Process: A machine was originally designed to handle sheets of metal separated by sheets of paper at very high speed. It is now being used for sheets of metal without the paper and it is causing unacceptable cosmetic damage to the sheets. Reintroducing the paper is not possible because of other processes downstream. Find a way to handle the sheets at high speed.
- Product: A food wrapper must prevent grease from penetrating, both for sanitary and aesthetic reasons. But, ink (which is very much like grease) must stick to the wrapper so that the product can be identified and advertised. How should the wrapper be constructed?
- Process: A system uses a highly purified closed loop water system and, as a result, utilizes a series of increasingly fine particulate filters. After yearly maintenance, which includes filter change outs, the system works great but slowly degrades with time as the finer filters become clogged and increase system pressure and, thus, reduces water flow rate. How can the operation be improved without simply adding more maintenance cycles, which of course reduces production time?
- Measurement: The customer specification requires that measurements be made during the cooling of the product. But, inserting the thermometer causes damage to the part of the product where the thermometer is inserted. Find a way to comply with the customer specification without wasting product.
- Process: A system produces a chemical product and a stream of waste material mixed with water. The waste has to be removed before the water can be recycled or disposed of. The present system requires three purification systems (one active, one being cleaned, one on standby) and an expensive, time-consuming method of cleaning (shovel the purifying material into a truck, transport it to a reprocessing facility, shovel it back into the truck, get it out of the truck and back into the system). Reduce the cost of the product by finding a way to make the waste purification system less expensive.

The flowchart (see Figure 15.1) shows several boxes with dashed lines, noted as "Consultant with your company." The champion will need to decide whether to use a consultant and, if the decision is "yes," will then need to select one. See Table 15.1 for a summary of the reasons to work with or without a consultant.

If you choose not to use a consultant, you can still use the flowchart shown in Figure 15.1. You will, however, need to get TRIZ training by other means, such as attending public seminars and conferences, reading TRIZ books and research papers, and such.[6]

If you decide to use a consultant, there are many resources available for finding one. Consultants and trainers populate the annual meetings of the Altshuller Institute and the European TRIZ Association.

Once you have decided on pilot projects and whether to use an outside consultant, you are ready to conduct the pilot project TRIZ class. We recommend an experiential style of teaching, in which the instructor teaches the basic principles and then coaches the class participants to solve real problems. This style has multiple benefits:

- Concepts for inventive solutions are generated for the selected projects or problems.
- The participants themselves generate the results.

Table 15.1 Should You Use a Consultant to Help Introduce TRIZ to Your Organization?

Yes	No
If you select a TRIZ training expert, it saves more time than it would take for your staff to learn TRIZ and develop training materials and methods.	Costs more than having a small number of employees study on their own.
Improves confidence of the pilot project participants because the consultant can show successful results from other organizations.	Some organizations reject anything that comes from outside, based on bad past experiences with consultants.
Produces more sophisticated results based on the consultant's experience. The beginners become advanced practitioners of TRIZ much more quickly than if they learn only from their own study and projects.	It may take time to educate the consultant about the company's culture and problems.

- The participants learn to sort the results and get immediate and long-term benefits.
- The participants learn the TRIZ methodology well enough to apply it themselves.

Implementation of the results of the pilot project is very important. Because some new problems might be generated during implementation, the class participants get to use their new skills and receive valuable reinforcement of what they have learned. The TRIZ results are visible to the organization, so that resistance to introduction of the new methods is reduced or eliminated.

After the successes in Step 2, the TRIZ champion and the organization's leadership pick one of two paths. The *internal* path produces a full team of internal TRIZ practitioners who replace the external instructors and consultants as their skills increase. The *external* path uses consultants to coach each team as each project is identified. The hybrid approach, in which the external path is followed for quick results while the internal path is followed for development of future self-sufficiency, has many benefits and is viewed by some as the best path. The external path is also frequently used for strategic planning, for applications of TRIZ to technology forecasting for the entire industry, and for product platforms during the time that it takes to develop the internal path. The internal experts learn the strategic uses of TRIZ and, during their advanced topics education, become internal consultants as well as instructors (see Figure 15.2)

The hybrid approach is very effective as it gains your company the best of both worlds: expert TRIZ training and project support from day one and the development of internal TRIZ expertise over time. As a matter of fact, this is the exact method employed by the author's company of employment when he was first exposed to TRIZ. The author worked for the world's largest semiconductor manufacturer, Intel Corporation, and was on the leading wave of external consultant–trained TRIZ employees. Over time, the author became the chair of Intel's worldwide TRIZ program, supporting the training of several thousand employees and overseeing projects with a combined impact of well over two billion US dollars. The external TRIZ training consultation group was completely eliminated from the internal TRIZ program somewhere around year four or five of the overall effort.

In other examples, one of the authors supported an almost identical transition from an external consultant–led TRIZ program to one supported entirely internally at both Dow Chemical

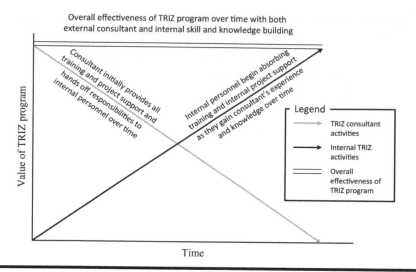

Overall effectiveness of TRIZ program over time with both external consultant and internal skill and knowledge building

Figure 15.2   Most effective method of TRIZ integration—hybrid approach.

and at Cummins. Further, it is known that General Electric, Siemens, Unilever, and Ford also followed the same pattern. Other worldwide industry players such as LG and Samsung have followed similar TRIZ journeys but with more reliance on deeper external consultation further into their innovation transformation.

Another example of the hybrid approach is the TRIZ implementation at the global lifting equipment producer, Konecranes Plc. The company started implementing TRIZ for product development in Finland in 2015. The training consisted of a 2-day learning period, which was followed by several days of student work. The primary goal for the training was to teach and deploy TRIZ into Konecranes while simultaneously achieving tangible results through the improvement of existing products and product development processes. Innovation specialist Ekku Rytkönen says that two days of intensive training followed by several days of individual practice gained real results.

15.4 Gaining Additional Benefits by Integrating TRIZ with Other Methodologies

15.4.1 Step 3

Step 3 is acceptance. As the organization develops its own internal experts, they take the lead in the integration of TRIZ with the organization's other methodologies. They become the collective champions in overcoming the last obstacle to TRIZ implementation: the traditional systems of project management. TRIZ will impact new product projects, process improvement projects, and process reengineering projects. Following the right-hand internal branch of the flowchart (Figure 15.1) will overcome this obstacle. As more and more people learn TRIZ and as the organization develops its internal cadre of experts, they will integrate TRIZ with all the company's other tools. Examples of the integration of TRIZ with existing tools include the following:

- TRIZ/QFD: Quality Function Deployment (QFD) identifies and prioritizes the *voice of the customer* and the capabilities of the organization's technologies, and then helps prioritize new concepts for design and production of products and services. TRIZ helps create the new concepts and resolve contradictions.[7–9]
- TRIZ/robust design/Taguchi methods: Robust design finds the right parameters to minimize all forms of waste and cost. TRIZ finds ways of creating the processes that will achieve those parameters.[10]
- TRIZ/DFM-A: Design for manufacturability and design for assembly identify and prioritize features of design that make manufacturing and assembly low cost and high yield with a short-cycle time. TRIZ resolves the technical problems encountered when implementing these features. Similarly, many organizations have developed their own guidelines for "design for serviceability," which is enhanced by TRIZ creativity in achieving serviceable designs.
- TRIZ/concurrent engineering (or integrated product and process engineering, or product development teams, or supplier/developer/customer teams): These project management teams will use TRIZ at many levels ranging from technology forecasting to conceptual design to production design and from implementation problem solving to service, delivery, and repair improvements.[11,12]
- TRIZ/Lean: Lean is a systematic method for identifying and eliminating seven types of waste from processes: Transport, Inventory, Motion, Waiting, Overproduction, Overprocessing, and Defects.[13] TRIZ is used with Lean to identify and solve contradictions associated with waste creation. An advanced usage of TRIZ, called TRIZ process analysis, has also been described as "Lean on steroids" and is exceedingly effective at resolving typical Lean issues.

The next chapter will show extended examples of the use of TRIZ with Six Sigma and other quality improvement methods.

At this level of integration, TRIZ passes from being seen as a tool, or a system of tools and methods, and becomes an intrinsic part of an organization's method of gaining competitive advantage and fulfilling customer needs. Until it reaches this point, it will require nurturing and championship to keep people aware of their opportunities to apply TRIZ.

15.5 Summary

The major steps for bringing TRIZ into an organization are recognition of the need for increased innovation, using TRIZ for pilot projects, and acceptance of TRIZ.

The organization's leadership will need to decide, depending on the organization's culture and on the time available, whether to use consultants or have their own employees do the entire process, or whether to use a hybrid of consultants and employees.

Use the worksheet in Table 15.2 to begin planning your TRIZ implementation.

Integration of TRIZ with other methods already in use in the organization, such as QFD, project management, design for manufacturability, and such, will accelerate its acceptance.

Most of the experiences and examples in this chapter are from medium-large to large companies. However, TRIZ works in any size company. For information on utilizing TRIZ within small- and medium-sized organizations, see the paper from the University of Bergamo, Italy.[14]

Table 15.2   A Worksheet for TRIZ Implementation

	Answer these questions to begin planning your TRIZ implementation.
1.	In my organization, who would be a good champion?
2.	Will that person need a higher-level management sponsor?
3.	If "yes," who would be a good sponsor?
4.	What will be the obstacles in my organization?
5.	What are the organization's strengths that TRIZ will increase?
6.	What are the organization's weaknesses that TRIZ will help overcome?
7.	Will we need to gather information about other organizations' successes with TRIZ to convince people that it can work in our company? If so, who will do the work of getting this information? (See Reference 2.)

References

1. Cowley, M. and Domb, E. 1997. *Beyond Strategic Vision: Effective Corporate Action with Hoshin Planning*. Boston, MA: Butterworth-Heinemann, chapter 2.
2. Spreafico, C. and Russo, D. 2016. TRIZ industrial case studies: A critical survey. *Procedia CIRP*, 39, 51–56. http://www.sciencedirect.com/science/article/pii/S2212827116001803.
3. Shaughnessy, H. 2013. What makes Samsung such an innovative company? *Forbes (Online) Magazine*, March 7. https://www.forbes.com/sites/haydnshaughnessy/2013/03/07/why-is-samsung-such-an-innovative-company/#5cd715c82ad7.

4. Yeo, H-S., Kim, J-H., and Lee, J-Y. 2015. Practice for solving problems and generating new concepts based on TRIZDAGEV roadmap. MATRIZ, TRIZfest, September 10–12, 164–173. http://matriz. org/wp-content/uploads/2012/07/TRIZfest-2015-conference-Proceedings.pdf .

5. JMP. 2015. Reducing cycles in the quest for clean energy. https://www.jmp.com/content/dam/jmp/ documents/en/customer-stories/nice/nice.pdf.

6. In the United States, the annual conference of the Altshuller Institute has tutorial sessions as well as research sessions. See www.aitriz.org. The European TRIZ Association holds an annual meeting that is primarily research oriented. Both meetings are good opportunities to learn TRIZ and to meet consultants and people from other companies to share experiences.

7. Schlueter, M. 2001. QFD by TRIZ. *The TRIZ Journal*, June 13. https://triz-journal.com/qfd-triz/.

8. Domb, E. and Corbin. D. 1998. QFD, TRIZ, and entrepreneurial intuition. The DelCor interactives international case study. *The TRIZ Journal*, September 5. https://triz-journal.com/ qfd-triz-entrepreneurial-intuition-delcor-interactives-international-case-study/.

9. León-Rovira, N. and Aguayo, H. 1998. A new model of the conceptual design process using QFD/FA/TRIZ. *The TRIZ Journal*, July 16. https://triz-journal.com/ new-model-conceptual-design-process-using-qfdfatriz/.

10. Phadke, M. (n.d.) Introduction to Robust Design (Taguchi Method). Retrieved from https://www.isixsigma.com/methodology/robust-design-taguchi-method/introduction-robust-design-taguchi-method/.

11. Cavallucci, D. and Lutz, P. 2000. Intuitive Design Method (IDM), a new approach on design methods integration. *The TRIZ Journal*, October 4. https://triz-journal.com/ intuitive-design-method-idm-new-approach-design-methods-integration/.

12. Zeidner, L. and Wood, R. 2000. The Collaborative Innovation (CI) process. *The TRIZ Journal*, June 3. https://triz-journal.com/collaborative-innovation-ci-process/.

13. Womack, J. P. and Jones, D. T. 2003. *Lean Thinking*. New York: Free Press, 352.

14. Russo, D., Regazzoni, D., and Rizzi, C. 2016. A long-term strategy to spread TRIZ in SMEs. Analysis of Bergamo's experience. Presented at ETRIA TRIZ Future Conference (TFC 2016), on October 25, 2016 at Wraclaw, Poland. http://www.osaka-gu.ac.jp/php/nakagawa/TRIZ/eTRIZ/epapers/e2017Papers/eRusso-LongTermStrategy/eRusso-TFC2016-LongTermStrategy-170211.html.

Chapter 16

Integrating TRIZ with Six Sigma and Other Quality Improvement Systems

16.1 Introduction

As time passes, ever more sophisticated methodologies are required to support the continued advancements in our technical and business environments. One method of "creating" more sophisticated methodologies is to combine separate proven and complimentary methods into "new" tool sets. This chapter discusses the powerful combinations of TRIZ with other popular analysis tools. Specifically, we will discuss the powerful combinations of TRIZ and Six Sigma (Section 16.2 and 16.3), TRIZ and QFD (Section 16.4), TRIZ and TOC (Section 16.5), and TRIZ and Lean (Section 16.6).

16.2 TRIZ with Six Sigma

The Six Sigma system for quality improvement in products, services, and processes is a business-based system of using statistical analysis and customer-focused methods. It has been demonstrated repeatedly that a company that moves from three sigma processes to six sigma processes increases its profitability by two to three orders of magnitude, and that companies that use the Design for Six Sigma (DFSS) process create products and services with much higher levels of customer satisfaction and technical quality than those that do not use DFSS.[1-6] Coupling TRIZ with Six Sigma produces these powerful results faster because the breakthrough problem-solving aspects of TRIZ can be focused on the profit opportunities identified by Six Sigma, and the technology-forecasting aspects of TRIZ can be focused on planning new products at the right time in the product life cycle.

At a high level, Six Sigma is very good at defining project requirements and boundary conditions and then also very good at driving implementation and testing for the success of the designed (or redesigned) system. However, in between the project planning and project implementation/

follow-up stages is a very crucial stage called "solution development." Unfortunately, when it comes time to create a solution, the guidance within Six Sigma is the same as every other non-TRIZ "problem-solving" process—good old-fashioned brainstorming. In other words, you are on your own. Therefore, TRIZ can be inserted into the Six Sigma process as the solution generation engine to augment and enhance the otherwise useful and effective Six Sigma process. We will now look in more detail as to how TRIZ can be inserted into Six Sigma with great benefit.

The breakthrough strategy of Six Sigma is different in vocabulary, but not in concept, from the *plan-do-check-act* method, usually known as "PDCA" (Plan, Do, Check, and Act), that has been used in quality improvement for the past 70 years and in human learning throughout our evolution.[7,8] The tools of TRIZ that are used in the improvement arena of Six Sigma are shown in Table 16.1 as well as the relationship to the PDCA model.[1] The difference in emphasis between Six Sigma and conventional quality improvement methods is the focus at all levels of Six Sigma application on the business results of the proposed improvement.

Special vocabulary is used for the roles of the people involved in Six Sigma. Typically, Six Sigma *champions* identify improvement projects, and *black belts* lead project teams to conduct the analysis and improvement, using the eight steps identified as A–H in Table 16.1, or perform the activities themselves. *Green belts* are members of the project teams, or leaders of teams, or may occasionally perform projects themselves, if the full skills of a black belt are not required. *Master black belts* train the black belts and green belts and serve as their advisors as they conduct their projects. In many companies, the black belts and master black belts are now getting TRIZ training as well as classical Six Sigma training so they can accelerate the improvement process.

Table 16.1 Opportunities to Apply TRIZ Occur in All of the Eight Phases of Six Sigma's Improvement Process (Sometimes Called "MAIC" after Phases C, D, E, and F)

Plan-Do-Check-Act Phase	*Six Sigma Phase*	*TRIZ Opportunities*
Plan	A. Recognize	Tool/object analysis, Ideal Final Result
Plan	B. Define	Tool/object analysis
Plan	C. Measure	Develop measurement methods, improve instruments using technology forecasting and contradiction analysis
Plan	D. Analyze	Contradiction analysis
Plan-Do-Check	E. Improve	Create new product, process, and service concepts (elimination of inherent or tradeoff contradictions, scientific effects)
Check-Act	F. Control	Same as C
Act	G. Standardize	Same as E, applied to service and product delivery system
Act, Plan	H. Integrate	Same as E, applied to whole system of improvement

Table 16.2   The Relationship between the TRIZ Tools and the Phases of Design for Six Sigma

DFSS Phase	TRIZ Tool
Multigenerational plan	Technology forecasting, tool/object analysis
Voice of customer and other elements of QFD	Conflict resolution, Ideal Final Result, development of measurement methods
Concept development	All
Detailed design	All
Optimize	Conflict resolution, trimming, problem solving
Validate/implement	Same

DFSS is used for either of two reasons:

1. To design new products, services, or processes that can function at Six Sigma quality level, or at whatever quality level is selected for business reasons, using the Six Sigma criteria.
2. To improve existing products, services, or processes if the improvement requires a discontinuous redesign from the earlier system. Improvement beyond 4.5 sigma (the so-called wall) often requires complete redesign.[5]

Typically, DFSS is merged with the company's previous methods of product development initially, and a Six Sigma methodology is developed after the company has extensive experience with pilot projects. Table 16.2 lists the phases of DFSS and the TRIZ tools that are useful in each phase.

16.3 Methods of Introducing TRIZ into Six Sigma

Six Sigma is a very highly structured system with a hierarchy of champions or project sponsors, master black belts, black belts, and green belts with defined levels of knowledge of business, identification of opportunities, statistical processes for analysis and control, and improvement at each level. Companies have inserted TRIZ into this process at a number of different points and in many different ways (training, workshop, consulting, etc.).

Motorola, the company that developed the Six Sigma process from its earlier quality improvement initiatives in the late 1980s, is at the least-structured end of the spectrum of methods of incorporating TRIZ into Six Sigma. TRIZ is taught and facilitated through the intellectual property organization. Six Sigma methods are taught and facilitated through a separate Six Sigma organization. Black belts frequently study TRIZ and use TRIZ methods to solve their problems, but no joint curriculum exists.

General Electric and AlliedSignal/Honeywell, the companies most famous for the economic benefits of their Six Sigma systems, have been similarly loosely structured. Many pockets of TRIZ knowledge exist within both companies that use a variety of TRIZ-derived methods and software systems and apply TRIZ to Six Sigma projects.

The Ford Motor Company has used TRIZ methods in a variety of ways since the early 1990s. They trained 400 people a year in the USIT (unified structured innovative thinking) version of TRIZ. In 2000–2001, Ford introduced TRIZ into the pilot project stage of their Six Sigma process.

Dow Chemical Company also piloted the use of TRIZ in both DFSS and the measure-ana-lyze-improve-control (MAIC [process improvement]) system in 2001–2002. More than 150 R&D staff, including 6 master black belts and 10 black belts received TRIZ training and applied it to their projects. For MAIC, a brief overview of TRIZ is presented to the black belts in their training, and they then decide whether to enroll in the TRIZ classes or to call on a TRIZ expert when their project team needs an innovative solution to a problem. For DFSS, the success of the pilot projects, which integrated QFD with TRIZ, led to the decision that all master black belts would be trained in TRIZ for problem solving and technology forecasting, as well as in the use of TRIZ-related software. They will then use TRIZ as needed and teach TRIZ to the black belts and to members of the DFSS teams. The progress through 2007 has been reported in *iSixSigma Magazine*.[9]

Delphi Automotive Systems is at the highly structured end of the spectrum of relationships between TRIZ and Six Sigma. TRIZ is used repeatedly in the design phase and in the process optimization phase. The overlap of tools and techniques between TRIZ, design of experiments, QFD, and other methods in the identification of the ideal system, function analysis, and iterative improvement is emphasized throughout the Delphi training program and the use of the DFSS process.[10]

16.4 TRIZ with QFD

TRIZ/QFD: QFD identifies and prioritizes the *voice of the customer* and the capabilities of the organization's technologies, and then helps prioritize new concepts for the design and production of products and services. TRIZ helps create the new concepts and resolve contradictions.

The relationship between TRIZ and QFD is best illustrated by the QFD matrix called the "house of quality."[11–13] After the QFD team has collected information by interviewing and observing the customer, the data is organized in a matrix, shown in Figure 16.1. Figure 16.2 indicates regions of the QFD house of quality matrix that signal the need for TRIZ. There are five obvious opportunities, marked on the matrix with an X, for interaction between QFD and TRIZ.

- ■ Box 8—Resolve conflict between performance measures.
- ■ Box 5—Empty rows. Use TRIZ to develop a means of satisfying customer needs.
- ■ Box 5—Empty columns. Use TRIZ to eliminate unnecessary activities. (Caution: Some actions may be necessary for regulatory reasons not obvious to the end user.)
- ■ Boxes 4 and 7—Use TRIZ to develop performance measures and measurement methods.

For example, when designing a house, customers might say they want it to feel spacious but to take very little time to clean. These would be "customer needs" in Box 1. The "performance measure" in Box 4 that corresponds to spaciousness might be the volume or the area of the rooms. This would be in conflict with making the house very fast to clean, assuming that is accomplished if that is done by making the rooms small. This would result in a contradiction notation in Box 8. TRIZ would then be used to resolve the contradiction. It could be treated as an inherent contradiction:

The room should be large (to feel spacious) but it should be small (to be quick to clean) or it could be treated as tradeoff contradiction:

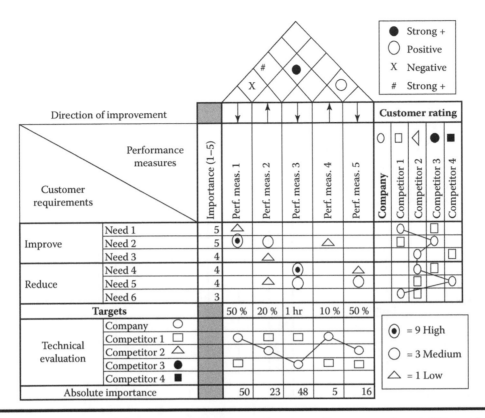

Figure 16.1 A typical QFD house of quality matrix. See Reference 9 to learn QFD methods. This matrix shows a strong correlation between Need 2 and Performance Measure 1, a medium correlation between Need 2 and Performance Measure 2, and a weak correlation between Need 2 and Performance Measure 4. There is strong conflict between Performance Measures 1 and 3 and positive reinforcement between Performance Measures 2 and 4.[11–13]

- IF the room is large
- THEN it will feel spacious
- BUT it will take longer to clean

Similarly, the other QFD matrices, such as the cost deployment matrix, the production (or service) planning matrix, the reliability matrix, and others, each have areas that will indicate, to those who are experienced in the use of QFD, the need for TRIZ.

16.5 Using TRIZ with the TOC

Eliyahu Goldratt introduced an integrated problem-solving tool set loosely known as the "Theory of Constraints" in the early 1990s. For those people who are already familiar with the TOC methods, this chapter is intended to show how TRIZ and TOC integrate very naturally. For those who would like more information on TOC, we suggest Goldratt's books,[14–16] H. William Dettmer's books,[17,18] and the articles in *The TRIZ Journal* by Domb and Dettmer[19] and Ed Moura.[20]

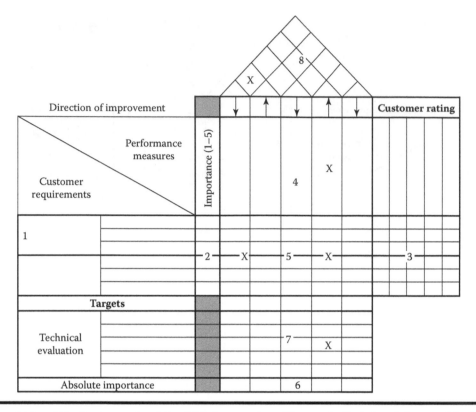

Figure 16.2   X indicates parts of the QFD house of quality matrix that signal the need for TRIZ.

The conflict resolution diagram (CRD), or "evaporating cloud," is one of the most powerful tools in the TOC tool set for resolving conflict. It is one of the few methods designed for formally structuring "win-win" solutions. In that respect, it is similar to TRIZ, in that both reject tradeoffs.

People who are familiar with TOC will recognize that the current reality tree and the CRD, either separately or together, show opportunities to use TRIZ to remove the cause of a problem.

16.6 Using TRIZ with Lean

TRIZ/Lean: Lean is a systematic method for identifying and eliminating seven types of waste from processes:

■ Transport (moving products that are not actually required to perform the processing).
■ Inventory (all components, work in process, and finished product not being processed).
■ Motion (people or equipment moving or walking more than is required to perform the processing).
■ Waiting (waiting for the next production step, interruptions of production during shift change).
■ Overproduction (production ahead of demand)
■ Overprocessing (resulting from poor tool or product design– creating activity).
■ Defects (the effort involved in inspecting for and fixing defects).[21]

Table 16.3   TRIZ Tools/Methods Support of Lean Steps

Lean Steps	Associated TRIZ Tools/Methods
Voice of customer	Root cause and functional analysis
Define value stream	Root cause, functional analysis, and Ideal Final Result statement
Identify waste	Contradiction and resource analysis
Develop solutions	40 principles and patterns of evolution
Eliminate waste	40 principles and patterns of evolution
Deliver product	Solution implementation
Seek perfection*	Evaluation for contradiction resolution, resource utilization, and Ideal Final Result movement

Note: The TRIZ tools for "seek perfection" should be executed before and after "deliver product."

While Lean is very good at identifying seven types of process waste, the methodology solution path is to simply "get rid of them." In some situations, the simple elimination of waste is easy. In most situations, it is not. Process designers do not foolishly design waste into their processes. There are usually contradictory requirements that result in waste. TRIZ is used with Lean to identify and solve contradictions associated with waste production. Further, Lean suggests that changes be made quickly and at the point of conflict. Unfortunately, this direction can result in localized changes that have negative impacts elsewhere in the system. The TRIZ full systems approach (partially supported by functional analysis) helps eliminate negative impacts to the larger system. Table 16.3 shows the relationship between the Lean steps and the TRIZ tools and methods that support those steps.

16.7 TRIZ Augments Most Every System

There are many other quality improvement initiatives in active use worldwide. Total quality management (TQM) evolved in the 1980s, was in wide use in the early 1990s, and emphasized the need for quality in the *total* business—planning, management, sales, service, employee relationships, product development, and so on—as well as in the production area that was usually the focus of quality efforts. TQM is now used widely in healthcare and education quality initiatives, as well as in business. ISO-9000 and the related standards QS-9000 and AS-9000 initially emphasized the need for standardization and documentation, but in their revisions in 1999 and 2000, they placed much more emphasis on understanding customer needs and on continuous improvement based on customer and technical data. Many companies and government agencies have quality initiatives without formal names—they have been committed to customer-focused, business-focused quality improvement for so long that it has become a part of the organization's culture, not a separate "quality thing."[8] TRIZ is helpful in all these processes. The obvious way to use TRIZ is to fix technical problems with products and services. Less obvious ways include the following:

■ Use TRIZ to find a creative way to get the customers' input. A classic problem for small companies doing international business is having no budget to travel to the customers' locations, to listen to and observe the customers. A problem for all companies in international business is lack of knowledge of their customers' language and culture. One Scandinavian electronics company found a TRIZ solution (using a *resource*) to the QFD challenge of listening to their female Japanese customers. They trained female Japanese employees of a subsidiary, who had done only production work, in customer interview skills. The results were excellent.[22]

■ Use TRIZ to resolve the conflicts between the customers' needs and the organization's traditional way of doing things. Many electronic business ideas for customer direct access to consulting firm databases are emerging directly from this research.[23]

■ Use TRIZ to resolve issues between the requirements of a new product or process and project boundary conditions or restrictions. Juan Aranda, an engineer at Intel Corporation who was trained in TRIZ alongside one of the the authors, created an elegant and almost free solution to a complex and expensive challenge. The HR department had a problem. Every time an employee moved to a new position within Intel, which happened frequently (the author himself had 11 different positions over his 17 years with the company), a new job code was assigned to that employee. The problem arose in that the HR software used to calculate salary and benefits did not allow for an employee's job code to be changed. Therefore, when an employee changed jobs and received a new job code, a new HR data file would have to be created for the employee, which required significant recalculation of salary and benefits, even though those usually did not change with the position. This led to a large amount of data entry, at the least, and reoccurring data errors at the worst. The quote from the software vendor to address the issue was in the hundreds of thousands of US dollars, of course an unacceptable expense to the HR department. Aranda used the "nine

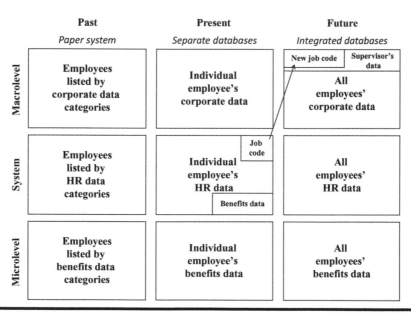

Figure 16.3   Employee job code change example.

screens" approach (see Section 3.5.3) to model the problem and search for a solution. To study the situation, he created the nine screens shown in Figure 16.3. As can be observed, the present-day system (Individual employee HR data in separate databases—center box) is where the employee's job code and benefits data was stored. Aranda realized that if he could disassociate the job code data from the employee's benefits data, but still keep the job code data associated with the employee, he could solve the problem. Along with job codes, all employees also have a supervisor. From studying the nine screens, he realized that he could move the job code data from the employee's data (Individual employee HR data—center box) to the supervisor's data (All employees' corporate data in integrated databases—top-right box), which already listed employees' names and employees' identification numbers. By moving the job data from the employees' data to the supervisors' data, Aranda disassociated the job codes from the rest of the employees' benefits data but still kept the job codes associated with the employees. Now, changes in job codes no longer forced a recalculation and re-entry of the benefits data.

If your organization has a successful structured improvement process such as Six Sigma, TOC, or any of the other improvement systems, the best way to introduce TRIZ may be as a family of tools that can help you resolve conflicts creatively. Once the organization has learned to appreciate TRIZ for problem solving, the expanded use of TRIZ for technology forecasting and strategic planning will be natural.

If your organization has no structured improvement process, the best way to introduce TRIZ is usually in each functional area as a problem-solving tool for that function. Once TRIZ is well established in certain key functions, it will spread to the rest of the company because it has proven its validity. Where to begin is a challenge that is specific to each company—we have seen successful implementations that started in manufacturing, engineering, customer service, sales, service development, warranty service, quality control, knowledge management, and intellectual property management, among others.

References

1. Domb, E. 2000. The role of TRIZ in Six Sigma management. Presented at TRIZCON2000, The Altshuller Institute, Nashua, New Hampshire, April 30–May 2, 2000.
2. Harry, M. and Schroeder, R. 2000. *Six Sigma*. New York: Doubleday.
3. Fisher, A. 2000. Rules for joining the cult of perfectability. *Fortune*, February 7, 206–208.
4. Harry, M. 2000. A new definition aims to connect quality with financial performance. *Quality Progress*, 33:1, 64–66.
5. Perez-Wilson, M. 1999. *Six Sigma*. Scottsdale, AZ: Advanced Systems Consultants.
6. Pande, P. S., Neuman, R. P., and Cavanagh, R. R. 2000. *The Six Sigma Way*. New York: McGraw-Hill.
7. Mann, D. 2000. Contradiction chains. *The TRIZ Journal*, January 14. https://triz-journal.com/contradiction-chains/.
8. Domb, E. 1999. Increase creativity to improve quality: TRIZ and the Baldrige Award Criteria. Presented at TRIZCON1999, The Altshuller Institute, Detroit, Michigan, March 7–9, 1999.
9. Reynard, S. 2007. Good chemistry: Dow Pairs six sigma and innovation. *iSixSigma Magazine*, 3:4, July/August, 20.
10. Brown, A., Jr. 2000. The role of robust engineering in innovation and continuous improvement methodologies. Presented at the ASI Six Sigma Symposium, Novi, Michigan, October 2000.
11. Schlueter, M. 2001. QFD by TRIZ. *The TRIZ Journal*, June 13. https://triz-journal.com/qfd-triz/.

12. Domb, E. and Corbin, D. 1998. QFD, TRIZ and entrepreneurial intuition the DelCor interactives international case study. *The TRIZ Journal*, September 5. https://triz-journal.com/qfd-triz-entrepreneurial-intuition-delcor-interactives-international-case-study.

13. León-Rovira, N. and Aguayo, H. 1998. A new model of the conceptual design process using QFD/FA/TRIZ. *The TRIZ Journal*, July 16. https://triz-journal.com/new-model-conceptual-design-process-using-qfdfatriz/.

14. Goldratt, E. M. and Cox, J. 1992. *The Goal*. Great Barrington, MA: North River Press.

15. Goldratt, E. M. 1994. *It's Not Luck*. Great Barrington, MA: North River Press.

16. Goldratt, E. M., Schragenheim, E., and Ptak, C. A. 2000. *Necessary but Not Sufficient*. Great Barrington, MA: North River Press.

17. Dettmer, H. W. 1996. *Goldratt's Theory of Constraints*. Milwaukee, WI: ASQ Quality Press.

18. Dettmer, H. W. 1998. *Breaking the Constraints to World-Class Performance*. Milwaukee, WI: ASQ Quality Press.

19. Domb, E. and Dettmer, H. W. 1999. Breakthrough innovation in conflict resolution. *The TRIZ Journal*, May 19. https://triz-journal.com/breakthrough-innovation-conflict-resolution-marrying-triz-thinking-process/.

20. Moura, E. C. 1999. TOC trees help TRIZ. *The TRIZ Journal*, September 20. https://triz-journal.com/toc-trees-help-triz/.

21. Womack, J. P. and Jones, D. T. 2003. *Lean Thinking*. New York: Free Press, 352.

22. Anecdote presented at 11th Annual Symposium of the Quality Function Deployment Institute, Novi, Michigan, June 2000.

23. Domb, E. and Mann, D. 2001. Using TRIZ to overcome business contradictions: Profitable e-commerce. *The TRIZ Journal*, April 20. https://triz-journal.com/using-triz-overcome-business-contradictions-profitable-e-commerce/. *Proceedings of the Portland International Conference on Managing Engineering Technology*, Portland State University Department of Engineering and Technology Management, Portland, Oregon, July 29–August 2, 2001, 15–21.

Chapter 17

Book Summary: Creative Problem Solving and Innovative Thinking in a Nutshell

A compact graphical model for problem solving was introduced in Chapter 2 and repeated many times. The model helps you learn and understand TRIZ so that you can use the methodology in situations when you do not have the book in hand. To remember the essential steps, the following list may help:

1. Model the system and the problem. Do not try to jump directly to the solution. Remember that you can download blank worksheets for modeling the problem and following the whole agenda for problem solving at http://www.innomationcorp.com/simplifiedTRIZ/.
2. Seek the contradictions behind the problem. Particularly, try to find one primary inherent contradiction. Ensure that the contradiction focuses on the location in the system where the problem occurs.
3. Map the resources of the system. Look for invisible resources. Do not be satisfied by easy solutions that require making the system more complex.
4. Formulate the Ideal Final Result. Do not be satisfied by a conventional compromise.
5. Check the solution against the criteria of ideality. Ask: "What is the primary contradiction? Is it solved without making the system more complex?"
6. Develop the solution further and improve it. It is cheaper and easier to repeat the solution cycle more than once than to try to implement a less ideal solution.
7. Use the patterns of evolution and principles for innovation to improve solutions and as independent tools to generate new ideas.
8. Integrate TRIZ with other methods already in use in the organization, such as QFD, TOC, Six Sigma, Lean, and so forth.

9. Look at TRIZ as an evolving theory, not as a rigid formula. Critically evaluate the TRIZ theory and methods. Improve your tools continuously—and publish what you have done, so that others can benefit.

Summarizing the implementation steps of TRIZ, we would like for you to also recall the background of TRIZ presented at the beginning of the book. If you suggest solutions that have many benefits and little to no cost, the immediate reaction may be that TRIZ is yet another overblown methodology. Try a simple exercise: recall one of your own personal bright moments in your career or business. You surely have examples of very good solutions. They may have been times when the customer has been especially happy and your organization has enjoyed a big success. On the other hand, perhaps you have managed to find a win-win solution to some difficult conflict between people so that everybody was satisfied. Maybe you remember a good engineering solution when the numbers of parts and operations were drastically decreased, performance improved, and costs were cut. Any solution to some problem that was important to you will serve as an example.

Now think about what made your big successes so different from other projects. Undoubtedly, your best solutions have provided big benefits with comparatively small costs and little to no harmful side effects. They have increased the ideality of the system, which improves when the increase in benefits outweighs any increase in the costs of, and the harm done by, the system. We have learned that there are solutions near ideality and that designing solutions that approach ideality is easier than initially suspected. Use your own examples to teach these concepts to others.

In the first chapter, we stressed that the point of this book is to learn how to create and recognize good solutions. Analyzing the common features of good solutions across industries, we can find tools for developing new ideas. We compared the tools with vehicles. A mediocre driver moves faster than the best runner. Other studies of creativity have mistakenly focused on the characteristics of people who are good problem solvers.

Discussions of creativity and innovation are often rather fruitless because only the improvement of people is considered. In Chapter 1, we referred to Theories X and Y, presented several decades ago by McGregor. Theory X says that, "People must be coerced, controlled, directed, threatened with punishment to get them to put forth adequate effort."[1] This assumption flourishes in stories of creative effort. You have heard the testimonials of differently abled people who have said that their difference (seen as a handicap by some) is what actually enabled them to succeed in wonderful ways. Popular as these stories may be, there are often others to the contrary. While there are stories of differently abled people who have "turned lemons into lemonade," there are also many stories of people who suffered without producing any extraordinary achievements. People get results *in spite of* accidents, persecution, or poverty, not *because of* hardships and disasters.

Equally popular is the belief that the main obstacle blocking innovation is the lack of resources and that pouring money into research and design will increase the output of innovation. Academic studies, such as several by Michael Porter and the most recent by Porter and Stern,[2] reinforce the belief that a rich environment, with capital, suppliers of components, and an innovation-supporting infrastructure are necessary. Certainly, this assumption has more appeal to common sense than the idea that accidents or poverty stimulate creativity. However, a systemic innovation process is needed.

Another comparison illustrates this. Literacy is a mental tool, widespread in developed countries. Conversely, one of the first things needed for economic development in countries with a large illiterate population is to give workers literacy—the mental vehicle. Money will not induce people to be more creative—they already have been very creative to survive and feed their families in poverty.

In a complex, highly developed society, the needs are less obvious, but a similar picture can be seen. Trying to improve creativity by giving people more time and money for simple brainstorming is like trying to turn poor runners into marathoners by giving them time to practice and offering money for winning races.

While recognizing that people differ in their natural ability to be creative, this book focus on the tools that everyone can use, regardless of their natural ability. You have learned to use a new tool, the law of increasing ideality, driven by the definition of the Ideal Final Result.

References

1. McGregor, D. 1960. *The Human Side of Enterprise*. New York: McGraw-Hill, 34.
2. Porter, M. A. and Stern, S. 2001. Innovation: Location matters. *MIT Sloan Management Review*, 42:4, 28–36.

Chapter 18

Get Started

TRIZ works. If you have done the exercises as you read this book, you know that TRIZ will help you find innovative solutions to problems, help you understand the evolution of systems, and help you develop more, and better, ideas faster.

Many other tools of TRIZ were not included in this book because our goal is to help you start using TRIZ quickly. Once you have mastered the TRIZ methods presented here, by applying them to real problems in your business and your personal life, you may want to learn more about advanced tools of TRIZ. However, our advice is not to try to learn more tools now. As the title of this chapter says, get started. As with the development of any new skill, the more you practice, the better and faster you will become at the application of TRIZ. Get going and have fun!

We invite our readers to send questions and comments to us at http://www.innomationcorp.com/simplifiedTRIZ/. Your comments will help us improve future editions of this book.

Glossary

One of the purposes of this book is to keep terminology as simple and exact as possible. New terms have replaced some older TRIZ words. For example, we speak of "tradeoff" instead of "technical or engineering contradiction." The first criterion for selecting terms is that they reflect the subject matter adequately and are compatible with everyday language and with professional language in industry.

Compatibility with old TRIZ terms has taken second place. Readers interested in the older terminology used in the TRIZ community should consult the glossaries prepared by Fey[1] and Savransky.[2]

action: the influence of one component on another, particularly the influence of the tool on the object, which results in a change (or maintenance) of the affected component

action (action principle): a function or a specific way of achieving a function

ARIZ: acronym for the Russian words *algorithm rezhenija izobretatelskih zadach*. An English translation is ASIP (algorithm for solving inventive problems). A step-by-step guide was developed by Altshuller for the analysis and resolving of contradictions. Altshuller and his team developed several versions of ARIZ between 1956 and 1985

ASIP: see **ARIZ**

auxiliary resource: resource that changes the **principal resources** so that the **inherent contradiction** is resolved

conflict: see **contradiction**

contradiction: opposition between things or properties of things. There are two kinds of contradictions: (1) **tradeoff** is a situation in which if something good happens, something bad also happens. Alternatively, if something good gets better, something undesirable gets worse. (2) **inherent contradiction** is a situation in which one thing needs two opposite properties

convolution (also see trimming): decreasing the number of parts and operations in the system but where the useful features and functions are retained. See **expansion**

engineering contradiction: see **tradeoff**

excessive function: a useful function that creates a desired change but to a level greater than wanted

expansion: increasing the number of parts and operations in the system so that useful features and functions are increased

feature: a property of the system

field: (also see **interaction**): an entity without rest mass that transmits interaction between substances. Examples include magnetic, electric, thermal, and acoustic fields

function: the term "function" is a diffused concept with many meanings: (1) Interaction including the **action** and the **object** of action (the motorcycle moves the person); (2) the purpose of the action (the purpose of the motorcycle may be to entertain the person); and (3) the result of the action (the motorcycle generates or produces noise and exhaust gas). More theoretically, an action performed by one material object (function carrier) to change or maintain a parameter of another material object (object of the function)

function model: a model of a function being performed. Performance of a useful function is measured as sufficient, insufficient or excessive

harmful function: a function that creates an undesired change

Ideal Final Result: (1) The solution that removes the **contradiction** using the **resources** in the system and its environment; (2) a description of the desired outcome, without use of jargon, that emphasizes achievement of the benefits of the system; (3) algebraically, the situation for the ideality equation when the denominator approaches zero: Ideality = Σ Benefits/(Σ Cost + Σ Harm)

inherent contradiction: a situation in which one thing has two opposite requirements. See also **contradiction** and **tradeoff**

insufficient function: a useful function that creates a desired change but to a level less than wanted

interaction: influence of the components of a system on each other. See **action**

instrument: see **tool**

model: an idealized concise description of phenomena and problems. A model contains relevant parts and connections between them and explains the relationship between those parts of the system

object: the component of a system that is influenced or acted on by the **tool**

object (of function): a material object, a parameter of which is changed as a result of a function being performed on it

parameter: a comparable value of an attribute. (color, speed, understanding, etc.)

pattern of evolution: a regularity discovered in the evolution of the system. Repetition of a sequence of similar events in the history of a system

physical contradiction: see **inherent contradiction**

principal resource: most important **resource** containing **inherent contradiction.** see **resources**

principle: principle for innovation. A generic solution applicable in many industries. The most widely used set of these solutions is the list of the 40 Principles

psychological inertia: the resistance to thinking in a new way. By analogy to physical inertia, thoughts continue in the same pattern unless disrupted by a force

resources: things, information, energy, time, space, or properties of the materials that are already in or near the environment of the system and are available for the resolution of the **contradiction** and achieving the **Ideal Final Result**

standard solution: typical transformation of the system, improving it and removing the **contradiction.** Altshuller and his team developed the 76 standard solutions

substance and field resources: see **resources**

sufficient function: a useful function that creates a desired change exactly to the level wanted

system: the set of objects and the interactions between them, having features or properties not reducible to the features or properties of separate objects. The system of objects is more than the sum of these objects because of the interactions between them. The set of interacting **tools** and **objects**

technical contradiction: See **tradeoff**

TIPS: see **TRIZ**

tool: component that influences or acts on the **object**

tradeoff: if something good happens, something bad also happens. Alternatively, if something good gets better, something undesirable gets worse

trimming: see **convolution**

TRIZ: theory of inventive problem solving **(TIPS)**. Acronym for the Russian words *teorija rezhenija izobretatelskih zadach*

useful function: a function that creates a desired change

value: the total functionality of a system divided by its total costs (which includes harmful)

zone of proximal development: solutions that are possible but have not yet been developed

References

1. Fey, V. 2000. TRIZ glossary. *Izobretenie*, II, 15.
2. Savransky, S. D. 2000. *Engineering of Creativity*. Boca Raton, FL: CRC Press.

Appendix

Features are defined in table 11.3. Same features can be used as Improving or Worsening depending on contradiction being analyzed

The Contradiction Matrix

		Worsened Feature									
		1	2	3	4	5	6	7	8	9	10
	Improved Feature	Weight of Moving Object	Weight of Stationary Object	Length of Moving Object	Length of Stationary Object	Area of Moving Object	Area of Stationary Object	Volume of Moving Object	Volume of Stationary Object	Speed	Force
1	Weight of moving object	All	All	15, 8, 29, 34	All	29, 17, 38, 34	All	29, 2, 40, 28	All	2, 8, 15, 38	8, 10, 18, 37
2	Weight of stationary object	All	All	All	10, 1, 29, 35	All	35, 30, 13, 2	All	5, 35, 14, 2	All	8, 10, 19, 35
3	Length of moving object	8, 15, 29, 34	All	All	All	15, 17, 4	All	7, 17, 4, 35	All	13, 4, 8	17, 10, 4
4	Length of stationary object	All	35, 28, 40, 29	All	All	All	17, 7, 10, 40	All	35, 8, 2, 14	All	28, 10
5	Area of moving object	2, 17, 29, 4	All	14, 15, 18, 4	All	All	All	7, 14, 17, 4	All	29, 30, 4, 34	19, 30, 35, 2
6	Area of stationary object	All	30, 2, 14, 18	All	26, 7, 9, 39	All	All	All	All	All	1, 18, 35, 36
7	Volume of moving object	2, 26, 29, 40	All	1, 7, 4, 35	All	1, 7, 4, 17	All	All	All	29, 4, 38, 34	15, 35, 36, 37
8	Volume of stationary object	All	35, 10, 19, 14	19, 14	35, 8, 2, 14	All	All	All	All	All	2, 18, 37
9	Speed	2, 28, 13, 38	All	13, 14, 8	All	29, 30, 34	All	7, 29, 34	All	All	13, 28, 15, 19
10	Force	8, 1, 37, 18	18, 13, 1, 28	17, 19, 9, 36	28, 10	19, 10, 15	1, 18, 36, 37	15, 9, 12, 37	2, 36, 18, 37	13, 28, 15, 12	All

(Continued)

The Contradiction Matrix (Continued)

	Improved Feature	Worsened Feature									
		1	2	3	4	5	6	7	8	9	10
		Weight of Moving Object	Weight of Stationary Object	Length of Moving Object	Length of Stationary Object	Area of Moving Object	Area of Stationary Object	Volume of Moving Object	Volume of Stationary Object	Speed	Force
11	Stress or pressure	10, 36, 37, 40	13, 29, 10, 18	35, 10, 36	35, 1, 14, 16	10, 15, 36, 28	10, 15, 36, 37	6, 35, 10	35, 24	6, 35, 36	36, 35, 21
12	Shape	8, 10, 29, 40	15, 10, 26, 3	29, 34, 5, 4	13, 14, 10, 7	5, 34, 4, 10	All	14, 4, 15, 22	7, 2, 35	35, 15, 34, 18	35, 10, 37, 40
13	Stability of object's composition	21, 35, 2, 39	26, 39, 1, 40	13, 15, 1, 28	37	2, 11, 13	39	28, 10, 19, 39	34, 28, 35, 40	33, 15, 28, 18	10, 35, 21, 16
14	Strength	1, 8, 40, 15	40, 26, 27, 1	1, 15, 8, 35	15, 14, 28, 26	3, 34, 40, 29	9, 40, 28	10, 15, 14, 7	9, 14, 17, 15	8, 13, 26, 14	10, 18, 3, 14
15	Duration of action by a moving object	19, 5, 34, 31	All	2, 19, 9	All	3, 17, 19	All	10, 2, 19, 30	All	3, 35, 5	19, 2, 16
16	Duration of action by stationary object	All	6, 27, 19, 16	All	1, 40, 35	All	All	All	35, 34, 38	All	All
17	Temperature	36, 22, 6, 38	22, 35, 32	15, 19, 9	15, 19, 9	3, 35, 39, 18	35, 38	34, 39, 40, 18	35, 6, 4	2, 28, 36, 30	35, 10, 3, 21
18	Illumination intensity	19, 1, 32	2, 35, 32	19, 32, 16	All	19, 32, 26	All	2, 13, 10	All	10, 13, 19	26, 19, 6
19	Use of energy by moving object	12, 18, 28, 31	All	12, 28	All	15, 19, 25	All	35, 13, 18	All	8, 35, 35	16, 26, 21, 2
20	Use of energy by stationary object	All	19, 9, 6, 27	All	All	All	All	All	All	All	36, 37

(Continued)

The Contradiction Matrix (Continued)

	Improved Feature	Worsened Feature									
		1	2	3	4	5	6	7	8	9	10
		Weight of Moving Object	Weight of Stationary Object	Length of Moving Object	Length of Stationary Object	Area of Moving Object	Area of Stationary Object	Volume of Moving Object	Volume of Stationary Object	Speed	Force
21	Power	8, 36, 38, 31	19, 26, 17, 27	1, 10, 35, 37		19, 38	17, 32, 13, 38	35, 6, 38	30, 6, 25	15, 35, 2	26, 2, 36, 35
22	Loss of energy	15, 6, 19, 28	19, 6, 18, 9	7, 2, 6, 13	6, 38, 7	15, 26, 17, 30	17, 7, 30, 18	7, 18, 23	7	16, 35, 38	36, 38
23	Loss of substance	35, 6, 23, 40	35, 6, 22, 32	14, 29, 10, 39	10, 28, 24	35, 2, 10, 31	10, 18, 39, 31	1, 29, 30, 36	3, 39, 18, 31	10, 13, 28, 38	14, 15, 18, 40
24	Loss of information	10, 24, 35	10, 35, 5	1, 26	26	30, 26	30, 16		2, 22	26, 32	
25	Loss of time	10, 20, 37, 35	10, 20, 26, 5	15, 2, 29	30, 24, 14, 5	26, 4, 5, 16	10, 35, 17, 4	2, 5, 34, 10	35, 16, 32, 18	All	10, 37, 36, 5
26	Quantity of substance/ matter	35, 6, 18, 31	27, 26, 18, 35	29, 14, 35, 18	All	15, 14, 29	2, 18, 40, 4	15, 20, 29	All	35, 29, 34, 28	35, 14, 3
27	Reliability	3, 8, 10, 40	3, 10, 8, 28	15, 9, 14, 4	15, 29, 28, 11	17, 10, 14, 16	32, 35, 40, 4	3, 10, 14, 24	2, 35, 24	21, 35, 11, 28	8, 28, 10, 3
28	Measurement accuracy	32, 35, 26, 28	28, 35, 25, 26	28, 26, 5, 16	32, 28, 3, 16	26, 28, 32, 3	26, 28, 32, 3	32, 13, 6	All	28, 13, 32, 24	32, 2
29	Manufacturing precision	28, 32, 13, 18	28, 35, 27, 9	10, 28, 29, 37	2, 32, 10	28, 33, 29, 32	2, 29, 18, 36	32, 23, 2	25, 10, 35	10, 28, 32	28, 19, 34, 36
30	External harm affects the object	22, 21, 27, 39	2, 22, 13, 24	17, 1, 39, 4	1, 18	22, 1, 33, 28	27, 2, 39, 35	22, 23, 37, 35	34, 39, 19, 27	21, 22, 35, 28	13, 35, 39, 18

(Continued)

The Contradiction Matrix (Continued)

	Improved Feature	Worsened Feature									
		1	2	3	4	5	6	7	8	9	10
		Weight of Moving Object	*Weight of Stationary Object*	*Length of Moving Object*	*Length of Stationary Object*	*Area of Moving Object*	*Area of Stationary Object*	*Volume of Moving Object*	*Volume of Stationary Object*	*Speed*	*Force*
31	Object-generated harmful factors	19, 22, 15, 39	35, 22, 1, 39	17, 15, 16, 22	All	17, 2, 18, 39	22, 1, 40	17, 2, 40	30, 18, 35, 4	35, 28, 3, 23	35, 28, 1, 40
32	Ease of manufacture	28, 29, 15, 16	1, 27, 36, 13	1, 29, 13, 17	15, 17, 27	13, 1, 26, 12	16, 40	13, 29, 1, 40	35	35, 13, 8, 1	35, 12
33	Ease of operation	25, 2, 13, 15	6, 13, 1, 25	1, 17, 13, 12	All	1, 17, 13, 16	18, 16, 15, 39	1, 16, 35, 15	4, 18, 39, 31	18, 13, 34	28, 13 35
34	Ease of repair	2, 27 35, 11	2, 27, 35, 11	1, 28, 10, 25	3, 18, 31	15, 13, 32	16, 25	25, 2, 35, 11	1	34, 9	1, 11, 10
35	Adaptability or versatility	1, 6, 15, 8	19, 15, 29, 16	35, 1, 29, 2	1, 35, 16	35, 30, 29, 7	15, 16	15, 35, 29	All	35, 10, 14	15, 17, 20
36	Device complexity	26, 30, 34, 36	2, 26, 35, 39	1, 19, 26, 24	26	14, 1, 13, 16	6, 36	34, 26, 6	1, 16	34, 10, 28	26, 16
37	Difficulty of detecting and measuring	27, 26, 28, 13	6, 13, 28, 1	16, 17, 26, 24	26	2, 13, 18, 17	2, 39, 30, 16	29, 1, 4, 16	2, 18, 26, 31	3, 4, 16, 35	30, 28, 40, 19
38	Extent of automation	28, 26, 18, 35	28, 26, 35, 10	14, 13, 17, 28	23	17, 14, 13	All	35, 13, 16	All	28, 10	2, 35
39	Productivity	35, 26, 24, 37	28, 27, 15, 3	18, 4, 28, 38	30, 7, 14, 26	10, 26, 34, 31	10, 35, 17, 7	2, 6, 34, 10	35, 37, 10, 2	All	28, 15, 10, 36

(Continued)

The Contradiction Matrix (Continued)

		Worsened Feature									
		11	12	13	14	15	16	17	18	19	20
	Improved Feature	*Stress or Pressure*	*Shape*	*Stability of Object's Composition*	*Strength*	*Duration of Action by a Moving Object*	*Duration of Action by Stationary Object*	*Temperature*	*Illumination Intensity*	*Use of Energy by Moving Object*	*Use of Energy by Stationary Object*
1	Weight of moving object	10, 36, 37, 40	10, 14, 35, 40	1, 35, 19, 39	28, 27, 18, 40	5, 34, 31, 35	All	6, 29, 4, 38	19, 1, 32	35, 12, 34, 31	All
2	Weight of stationary object	13, 29, 10, 18	13, 10, 29, 14	26, 39, 1, 40	28, 2, 10, 27	All	2, 27, 19, 6	28, 19, 32, 22	19, 32, 35	All	18, 19, 28, 1
3	Length of moving object	1, 8, 35	1, 8, 10, 29	1, 8, 15, 34	8, 35, 29, 34	19	All	10, 15, 19	32	8, 35, 24	All
4	Length of stationary object	1, 14, 35	13, 14, 15, 7	39, 37, 35	15, 14, 28, 26	All	1, 10, 35	3, 35, 38, 18	3, 25	All	All
5	Area of moving object	10, 15, 36, 28	5, 34, 29, 4	11, 2, 13, 39	3, 15, 40, 14	6, 3	All	2, 15, 16	15, 32, 19, 13	19, 32	All
6	Area of stationary object	10, 15, 36, 37	All	2, 38	40	All	2, 10, 19, 30	35, 39, 38	All	All	All
7	Volume of moving object	6, 35, 36, 37	1, 15, 29, 4	28, 10, 1, 39	9, 14, 15, 7	6, 35, 4	All	34, 39, 10, 18	2, 13, 10	35	All
8	Volume of stationary object	24, 35	7, 2, 35	34, 28, 35, 40	9, 14, 17, 15	All	35, 34, 38	35, 6, 4	All	All	All
9	Speed	6, 18, 38, 40	35, 15, 18, 34	28, 33, 1, 18	8, 3, 26, 14	3, 19, 35, 5	All	28, 30, 36, 2	10, 13, 19	8, 15, 35, 38	All
10	Force	18, 21, 11	10, 35, 40, 34	35, 10, 21	35, 10, 14, 27	19, 2	All	35, 10, 21	All	19, 17, 10	1, 16, 36, 37

(Continued)

The Contradiction Matrix (Continued)

	Improved Feature	Worsened Feature									
		11	12	13	14	15	16	17	18	19	20
		Stress or Pressure	Shape	Stability of Object's Composition	Strength	Duration of Action by a Moving Object	Duration of Action by Stationary Object	Temperature	Illumination Intensity	Use of Energy by Moving Object	Use of Energy by Stationary Object
11	Stress or pressure	All	35, 4, 15, 10	35, 33, 2, 40	9, 18, 3, 40	19, 3, 27	All	35, 39, 19, 2	All	14, 24, 10, 37	All
12	Shape	34, 15, 10, 14	All	33, 1, 18, 4	30, 14, 10, 40	14, 26, 9, 25	All	22, 14, 19, 32	13, 15, 32	2, 6, 34, 14	All
13	Stability of object's composition	2, 35, 40	22, 1, 18, 4	All	17, 9, 15	13, 27, 10, 35	39, 3, 35, 23	35, 1, 32	32, 3, 27, 16	13, 19	27, 4, 29, 18
14	Strength	10, 3, 18, 40	10, 30, 35, 40	13, 17, 35	v	27, 3, 26	All	30, 10, 40	35, 19	19, 35, 10	35
15	Duration of action by a moving object	19, 3, 27	14, 26, 28, 25	13, 3, 35	27, 3, 10	All	All	19, 35, 39	2, 19, 4, 35	28, 6, 35, 18	All
16	Duration of action by stationary object	All	All	39, 3, 35, 23	All	All	All	19, 18, 36, 40	All	All	All
17	Temperature	35, 39, 19, 2	14, 22, 19, 32	1, 35, 32	10, 30, 22, 40	19, 13, 39	19, 18, 36, 40	All	32, 30, 21, 16	19, 15, 3, 17	All
18	Illumination intensity	All	32, 30	32, 3, 27	35, 19	2, 19, 6	All	32, 35, 19	All	32, 1, 19	32, 35, 1, 15
19	Use of energy by moving object	23, 14, 25	12, 2, 29	19, 13, 17, 24	5, 19, 9, 35	28, 35, 6, 18	All	19, 24, 3, 14	2, 15, 19	All	All
20	Use of energy by stationary object	All	All	27, 4, 29, 18	35	All	All	All	19, 2, 35, 32	All	All

(Continued)

The Contradiction Matrix (Continued)

		Worsened Feature									
		11	12	13	14	15	16	17	18	19	20
	Improved Feature	*Stress or Pressure*	*Shape*	*Stability of Object's Composition*	*Strength*	*Duration of Action by a Moving Object*	*Duration of Action by Stationary Object*	*Temperature*	*Illumination Intensity*	*Use of Energy by Moving Object*	*Use of Energy by Stationary Object*
21	Power	22, 10, 35	29, 14, 2, 40	35, 32, 15, 31	26, 10, 28	19, 35, 10, 38	16	2, 14, 17, 25	16, 6, 19	16, 6, 19, 37	All
22	Loss of energy	All	All	14, 2, 39, 6	26	All	All	19, 38, 7	1, 13, 32, 15	All	All
23	Loss of substance	3, 36, 37, 10	29, 35, 3, 5	2, 14, 30, 40	35, 28, 31, 40	28, 27, 3, 18	27, 16, 18, 38	21, 36, 39, 31	1, 6, 13	35, 18, 24, 5	28, 27, 12, 31
24	Loss of information	All	All	All	All	10	10	All	19	All	All
25	Loss of time	37, 36, 4	4, 10, 34, 17	35, 3, 22, 5	29, 3, 28, 18	20, 10, 28, 18	28, 20, 10, 16	35, 29, 21, 18	1, 19, 26, 17	35, 38, 19, 18	1
26	Quantity of substance/ matter	10, 36, 14, 3	35, 14	15, 2, 17, 40	14, 35, 34, 10	3, 35, 10, 40	3, 35, 31	3, 17, 39	All	34, 29, 16, 18	3, 35, 31
27	Reliability	10, 24, 35, 19	35, 1, 16, 11	All	11, 28	2, 35, 3, 25	34, 27, 6, 40	3, 35, 10	11, 32, 13	21, 11, 27, 19	36, 23
28	Measurement accuracy	6, 28, 32	6, 28, 32	32, 35, 13	28, 6, 32	28, 6, 32	10, 26, 24	6, 19, 28, 24	6, 1, 32	3, 6, 32	All
29	Manufacturing precision	3, 35	32, 30, 40	30, 18	3, 27	3, 27, 40	All	19, 26	3, 32	32, 2	All
30	External harm affects the object	22, 2, 37	22, 1, 3, 35	35, 24, 30, 18	18, 35, 37, 1	22, 15, 33, 28	17, 1, 40, 33	22, 33, 35, 2	1, 19, 32, 13	1, 24, 6, 27	10, 2, 22, 37

(Continued)

The Contradiction Matrix (Continued)

		Worsened Feature									
		11	12	13	14	15	16	17	18	19	20
	Improved Feature	Stress or Pressure	Shape	Stability of Object's Composition	Strength	Duration of Action by a Moving Object	Duration of Action by Stationary Object	Temperature	Illumination Intensity	Use of Energy by Moving Object	Use of Energy by Stationary Object
31	Object-generated harmful factors	2, 33, 27, 18	35, 1	35, 40, 27, 39	15, 35, 22, 2	15, 22, 33, 31	21, 39, 16, 22	22, 35, 2, 24	19, 24, 39, 32	2, 35, 6	19, 22, 18
32	Ease of manufacture	35, 19, 1, 37	1, 28, 13, 27	11, 13, 1	1, 3, 10, 32	27, 1, 4	35, 16	27, 26, 18	28, 24, 27, 1	28, 26, 27, 1	1, 4
33	Ease of operation	2, 32, 12	15, 34, 29, 28	32, 35, 30	32, 40, 3, 28	29, 3, 8, 25	1, 16, 25	26, 27, 13	13, 17, 1, 24	1, 13, 24	All
34	Ease of repair	13	1, 13, 2, 4	2, 35	11, 1, 2, 9	11, 29, 28, 27	1	4, 10	15, 1, 13	15, 1, 28, 16	All
35	Adaptability or versatility	35, 16	15, 37, 1, 8	35, 30, 14	35, 3, 32, 6	13, 1, 35	2, 16	27, 2, 3, 35	6, 22, 26, 1	19, 35, 29, 13	All
36	Device complexity	19, 1, 35	29, 13, 28, 15	2, 22, 17, 19	2, 13, 28	10, 4, 28, 15	All	2, 17, 13	24, 17, 13	27, 2, 29, 28	All
37	Difficulty of detecting and measuring	35, 36, 37, 32	27, 13, 1, 39	11, 22, 39, 30	27, 3, 15, 28	19, 29, 39, 25	25, 34, 6, 35	3, 27, 35, 16	2, 24, 26	35, 38	19, 35, 16
38	Extent of automation	13, 35	15, 32, 1, 13	18, 1	25, 13	6, 9	All	26, 2, 19	8, 32, 19	2, 32, 13	All
39	Productivity	10, 37, 14	14, 10, 34, 40	35, 3, 22, 39	29, 28, 10, 18	35, 10, 2, 18	20, 10, 16, 38	35, 21, 28, 10	26, 17, 19, 1	35, 10, 38, 19	1

(Continued)

The Contradiction Matrix (Continued)

					Worsened Feature						
		21	22	23	24	25	26	27	28	29	30
	Improved Feature	*Power*	*Loss of Energy*	*Loss of Substance*	*Loss of Information*	*Loss of Time*	*Quantity of Substance/ Matter*	*Reliability*	*Measurement Accuracy*	*Manufacturing Precision*	*External Harm Affects the Object*
1	Weight of moving object	12, 36, 18, 31	6, 2, 34, 19	5, 35, 3, 31	10, 24, 35	10, 35, 20, 28	3, 26, 18, 31	1, 3, 11, 27	28, 27, 35, 26	28, 35, 26, 18	22, 21, 18, 27
2	Weight of stationary object	15, 19, 18, 22	18, 19, 28, 15	5, 8, 13, 30	10, 15, 35	10, 20, 35, 26	19, 6, 18, 26	10, 28, 8, 3	18, 26, 28	10, 1, 35, 17	2, 19, 22, 37
3	Length of moving object	1, 35	7, 2, 35, 39	4, 29, 23, 10	1, 24	15, 2, 29	29, 35	10, 14, 29, 40	28, 32, 4	10, 28, 29, 37	1, 15, 17, 24
4	Length of stationary object	12, 8	6, 28	10, 28, 24, 35	24, 26,	30, 29, 14	All	15, 29, 28	32, 28, 3	2, 32, 10	1, 18
5	Area of moving object	19, 10, 32, 18	15, 17, 30, 26	10, 35, 2, 39	30, 26	26, 4	29, 30, 6, 13	29, 9	26, 28, 32, 3	2, 32	22, 33, 28, 1
6	Area of stationary object	17, 32	17, 7, 30	10, 14, 18, 39	30, 16	10, 35, 4, 18	2, 18, 40, 4	32, 35, 40, 4	26, 28, 32, 3	2, 29, 18, 36	27, 2, 39, 35
7	Volume of moving object	35, 6, 13, 18	7, 15, 13, 16	36, 39, 34, 10	2, 22	2, 6, 34, 10	29, 30, 7	14, 1, 40, 11	25, 26, 28	25, 28, 2, 16	22, 21, 27, 35
8	Volume of stationary object	30, 6	All	10, 39, 35, 34	All	35, 16, 32, 18	35, 3	2, 35, 16	All	35, 10, 25	34, 39, 19, 27
9	Speed	19, 35, 38, 2	14, 20, 19, 35	10, 13, 28, 38	13, 26	All	10, 19, 29, 38	11, 35, 27, 28	28, 32, 1, 24	10, 28, 32, 25	1, 28, 35, 23
10	Force	19, 35, 18, 37	14, 15	8, 35, 40, 5	All	10, 37, 36	14, 29, 18, 36	3, 35, 13, 21	35, 10, 23, 24	28, 29, 37, 36	1, 35, 40, 18

(Continued)

The Contradiction Matrix (Continued)

	Improved Feature	Worsened Feature									
		21	22	23	24	25	26	27	28	29	30
		Power	*Loss of Energy*	*Loss of Substance*	*Loss of Information*	*Loss of Time*	*Quantity of Substance/ Matter*	*Reliability*	*Measurement Accuracy*	*Manufacturing Precision*	*External Harm Affects the Object*
11	Stress or pressure	10, 35, 14	2, 36, 25	10, 36, 3, 37	All	37, 36, 4	10, 14, 36	10, 13, 19, 35	6, 28, 25	3, 35	22, 2, 37
12	Shape	4, 6, 2	14	35, 29, 3, 5	All	14, 10, 34, 17	36, 22	10, 40, 16	28, 32, 1	32, 30, 40	22, 1, 2, 35
13	Stability of object's composition	32, 35, 27, 31	14, 2, 39, 6	2, 14, 30, 40	All	35, 27	15, 32, 35	All	13	18	35, 24, 30, 18
14	Strength	10, 26, 35, 28	35	35, 28, 31, 40	All	29, 3, 28, 10	29, 10, 27	11, 3	3, 27, 16	3, 27	18, 35, 37, 1
15	Duration of action by a moving object	19, 10, 35, 38	All	28, 27, 3, 18	10	20, 10, 28, 18	3, 35, 10, 40	11, 2, 13	3	3, 27, 16, 40	22, 15, 33, 28
16	Duration of action by stationary object	16	All	27, 16, 18, 38	10	28, 20, 10, 16	3, 35, 31	34, 27, 6, 40	10, 26, 24	All	17, 1, 40, 33
17	Temperature	2, 14, 17, 25	21, 17, 35, 38	21, 36, 29, 31	All	35, 28, 21, 18	3, 17, 30, 39	19, 35, 3, 10	32, 19, 24	24	22, 33, 35, 2
18	Illumination intensity	32	13, 16, 1, 6	13, 1	1, 6	19, 1, 26, 17	1, 19	All	11, 15, 32	3, 32	15, 19
19	Use of energy by moving object	6, 19, 37, 18	12, 22, 15, 24	35, 24, 18, 5	All	35, 38, 19, 18	34, 23, 16, 18	19, 21, 11, 27	3, 1, 32	All	1, 35, 6, 27
20	Use of energy by stationary object	All	All	28, 27, 18, 31	All	All	3, 35, 31	10, 36, 23	All	All	10, 2, 22, 37

(Continued)

The Contradiction Matrix (Continued)

		Worsened Feature									
		21	*22*	*23*	*24*	*25*	*26*	*27*	*28*	*29*	*30*
	Improved Feature	*Power*	*Loss of Energy*	*Loss of Substance*	*Loss of Information*	*Loss of Time*	*Quantity of Substance/ Matter*	*Reliability*	*Measurement Accuracy*	*Manufacturing Precision*	*External Harm Affects the Object*
21	Power	All	10, 35, 38	28, 27, 18, 38	10, 19	35, 20, 10, 6	4, 34, 19	19, 24, 26, 31	32, 15, 2	32, 2	19, 22, 31, 2
22	Loss of energy	3, 38	All	35, 27, 2, 37	19, 10	10, 18, 32, 7	7, 18, 25	11, 10, 35	32	All	21, 22, 35, 2
23	Loss of substance	28, 27, 18, 38	35, 27, 2, 31	All	All	15, 18, 35, 10	6, 3, 10, 24	10, 29, 39, 35	16, 34, 31, 28	35, 10, 24, 31	33, 22, 30, 40
24	Loss of information	10, 19	19, 10	All	All	24, 26, 28, 32	24, 28, 35	10, 28, 23	All	All	22, 10, 1
25	Loss of time	35, 20, 10, 6	10, 5, 18, 32	35, 18, 10, 39	24, 26, 28, 32	All	35, 38, 18, 16	10, 30, 4	24, 34, 28, 32	24, 26, 28, 18	35, 18, 34
26	Quantity of substance/ matter	35	7, 18, 25	6, 3, 10, 24	24, 28, 35	35, 38, 18, 16	All	18, 3, 28, 40	13, 2, 28	33, 30	35, 33, 29, 31
27	Reliability	21, 11, 26, 31	10, 11, 35	10, 35, 29, 39	10, 28	10, 30, 4	21, 28, 40, 3	All	32, 3, 11, 23	11, 32, 1	27, 35, 2, 40
28	Measurement accuracy	3, 6, 32	26, 32, 27	10, 16, 31, 28	All	24, 34, 28, 32	2, 6, 32	5, 11, 1, 23	All	All	28, 24, 22, 26
29	Manufacturing precision	32, 2	13, 32, 2	35, 31, 10, 24	All	32, 26, 28, 18	32, 30	11, 32, 1	All	All	26, 28, 10, 36
30	External harm affects the object	19, 22, 31, 2	21, 22, 35, 2	33, 22, 19, 40	22, 10, 2	35, 18, 34	35, 33, 29, 31	27, 24, 2, 40	28, 33, 23, 26	26, 28, 10, 18	All

(Continued)

The Contradiction Matrix (Continued)

	Improved Feature	Worsened Feature									
		21 Power	22 Loss of Energy	23 Loss of Substance	24 Loss of Information	25 Loss of Time	26 Quantity of Substance/ Matter	27 Reliability	28 Measurement Accuracy	29 Manufacturing Precision	30 External Harm Affects the Object
31	Object-generated harmful factors	2, 35, 18	21, 35, 2, 22	10, 1, 34	10, 21, 29	1, 22	3, 24, 39, 1	24, 2, 40, 39	3, 33, 26	4, 17, 34, 26	All
32	Ease of manufacture	27, 1, 12, 24	19, 35	15, 34, 33	32, 24, 18, 16	35, 28, 34, 4	35, 23, 1, 24		1, 35, 12, 18	All	24, 2
33	Ease of operation	35, 34, 2, 10	2, 19, 13	28, 32, 2, 24	4, 10, 27, 22	4, 28, 10, 34	12, 35	17, 27, 8, 40	25, 13, 2, 34	1, 32, 35, 23	2, 25, 28, 39
34	Ease of repair	15, 10, 32, 2	15, 1, 32, 19	2, 35, 34, 27	All	32, 1, 10, 25	2, 28, 10, 25	11, 10, 1, 16	10, 2, 13	25, 10	35, 10, 2, 16
35	Adaptability or versatility	19, 1, 29	18, 15, 1	15, 10, 2, 13	All	35, 28	3, 35, 15	35, 13, 8, 24	35, 5, 1, 10	All	35, 11, 32, 31
36	Device complexity	20, 19, 30, 34	10, 35, 13, 2	35, 10, 28, 29	All	6, 29	13, 3, 27, 10	13, 35, 1	2, 26, 10, 34	26, 24, 32	22, 19, 29, 40
37	Difficulty of detecting and measuring	18, 1, 16, 10	35, 3, 15, 19	1, 18, 10, 24	35, 33, 27, 22	18, 28, 32, 9	3, 27, 29, 18	27, 40, 28, 8	26, 24, 32, 28	All	22, 19, 29, 28
38	Extent of automation	28, 2, 27	23, 28	35, 10, 18, 5	35, 33	24, 28, 35, 30	35, 13	11, 27, 32	28, 26, 10, 34	28, 26, 18, 23	2, 33
39	Productivity	35, 20, 10	28, 10, 29, 35	28, 10, 35, 23	13, 15, 23	All	35, 38	1, 35, 10, 38	1, 10, 34, 28	18, 10, 32, 1	22, 35, 13, 24

(Continued)

The Contradiction Matrix (Continued)

	Improved Feature	Worsened Feature								
		31	32	33	34	35	36	37	38	39
		Object Generated Harmful Factors	Ease of Manufacture	Ease of Operation	Ease of Repair	Adaptability or Versatility	Device Complexity	Difficulty of Detecting and Measuring	Extent of Automation	Productivity
1	Weight of moving object	22, 35, 31, 39	27, 28, 1, 36	35, 3, 2, 24	2, 27, 28, 11	29, 5, 15, 8	26, 30, 36, 34	28, 29, 26, 32	26, 35, 18, 19	35, 3, 24, 37
2	Weight of stationary object	35, 22, 1, 39	28, 1, 9	6, 13, 1, 32	2, 27, 28, 11	19, 15, 29	1, 10, 26, 39	25, 28, 17, 15	2, 26, 35	1, 28, 15, 35
3	Length of moving object	17, 15	1, 29, 17	15, 29, 35, 4	1, 28, 10	14, 15, 1, 16	1, 19, 26, 24	35, 1, 26, 24	17, 24, 26, 16	14, 4, 28, 29
4	Length of stationary object	All	15, 17, 27	2, 25	3	1, 35	1, 26	26	All	30, 14, 7, 26
5	Area of moving object	17, 2, 18, 39	13, 1, 26, 24	15, 17, 13, 16	15, 13, 10, 1	15, 30	14, 1, 13	2, 36, 26, 18	14, 30, 28, 23	10, 26, 34, 2
6	Area of stationary object	22, 1, 40	40, 16	16, 4	16	15, 16	1, 18, 36	2, 35, 30, 18	23	10, 15, 17, 7
7	Volume of moving object	17, 2, 40, 1	29, 1, 40	15, 13, 30, 12	10	15, 29	26, 1	29, 26, 4	35, 34, 16, 24	10, 6, 2, 34
8	Volume of stationary object	30, 18, 35, 4	35	All	1	All	1, 31	2, 17, 26		35, 37, 10, 2
9	Speed	2, 24, 35, 21	35, 13, 8, 1	32, 28, 13, 12	34, 2, 28, 27	15, 10, 26	10, 28, 4, 34	3, 34, 27, 16	10, 18	All
10	Force	13, 3, 36, 24	15, 37, 18, 1	1, 28, 3, 25	15, 1, 11	15, 17, 18, 20	26, 35, 10, 18	36, 37, 10, 19	2, 35	3, 28, 35, 37

(Continued)

The Contradiction Matrix (Continued)

	Improved Feature	Worsened Feature									
		31 Object Generated Harmful Factors	32 Ease of Manufacture	33 Ease of Operation	34 Ease of Repair	35 Adaptability or Versatility	36 Device Complexity	37 Difficulty of Detecting and Measuring	38 Extent of Automation	39 Productivity	
11	Stress or pressure	2, 33, 27, 18	1, 35, 16	11	2	35	19, 1, 35	2, 36, 37	35, 24	10, 14, 35, 37	
12	Shape	35, 1	1, 32, 17, 28	32, 15, 26	2, 13, 1	1, 15, 29	16, 29, 1, 28	15, 13, 39	15, 1, 32	17, 26, 34, 10	
13	Stability of object's composition	35, 40, 27, 39	35, 19	32, 35, 30	2, 35, 10, 16	35, 30, 34, 2	2, 35, 22, 26	35, 22, 39, 23	1, 8, 35	23, 35, 40, 3	
14	Strength	15, 35, 22, 2	11, 3, 10, 32	32, 40, 25, 2	27, 11, 3	15, 3, 32	2, 13, 25, 28	27, 3, 15, 40	15	29, 35, 10, 14	
15	Duration of action by a moving object	21, 39, 16, 22	27, 1, 4	12, 27	29, 10, 27	1, 35, 13	10, 4, 29, 15	19, 29, 39, 35	6, 10	35, 17, 14, 19	
16	Duration of action by stationary object	22	35, 10	1	1	2	All	25, 34, 6, 35	1	20, 10, 16, 38	
17	Temperature	22, 35, 2, 24	26, 27	26, 27	4, 10, 16	2, 18, 27	2, 17, 16	3, 27, 35, 31	26, 2, 19, 16	15, 28, 35	
18	Illumination intensity	35, 19, 32, 39	19, 35, 28, 26	28, 26, 19	15, 17, 13, 16	15, 1, 19	6, 32, 13	32, 15	2, 26, 10	2, 25, 16	
19	Use of energy by moving object	2, 35, 6	28, 26, 30	19, 35	1, 15, 17, 28	15, 17, 13, 16	2, 29, 27, 28	35, 38	32, 2	12, 28, 35	
20	Use of energy by stationary object	19, 22, 18	1, 4	All	All	All	All	19, 35, 16, 25	All	1, 6	

(Continued)

The Contradiction Matrix (Continued)

		Worsened Feature								
		31	32	33	34	35	36	37	38	39
	Improved Feature	*Object Generated Harmful Factors*	*Ease of Manufacture*	*Ease of Operation*	*Ease of Repair*	*Adaptability or Versatility*	*Device Complexity*	*Difficulty of Detecting and Measuring*	*Extent of Automation*	*Productivity*
21	Power	2, 35, 18	26, 10, 34	26, 35, 10	35, 2, 10, 34	19, 17, 34	20, 19, 30, 34	19, 35, 16	28, 2, 17	28, 35, 34
22	Loss of energy	21, 35, 2, 22	All	35, 32, 1	2, 19	All	7, 23	35, 3, 15, 23	2	28, 10, 29, 35
23	Loss of substance	10, 1, 34, 29	15, 34, 33	32, 28, 2, 24	2, 35, 34, 27	15, 10, 2	35, 10, 28, 24	35, 18, 10, 13	35, 10, 18	28, 35, 10, 23
24	Loss of information	10, 21, 22	32	27, 22	All	All	All	35, 33	35	13, 23, 15
25	Loss of time	35, 22, 18, 39	35, 28, 34, 4	4, 28, 10, 34	32, 1, 10	35, 28	6, 29	18, 28, 32, 10	24, 28, 35, 30	All
26	Quantity of substance/ matter	3, 35, 40, 39	29, 1, 35, 27	35, 29, 25, 10	2, 32, 10, 25	15, 3, 29	3, 13, 27, 10	3, 27, 29, 18	8, 35	13, 29, 3, 27
27	Reliability	35, 2, 40, 26	All	27, 17, 40	1, 11	13, 35, 8, 24	13, 35, 1	27, 40, 28	11, 13, 27	1, 35, 29, 38
28	Measurement accuracy	3, 33, 39, 10	6, 35, 25, 18	1, 13, 17, 34	1, 32, 13, 11	13, 35, 2	27, 35, 10, 34	26, 24, 32, 28	28, 2, 10, 34	10, 34, 28, 32
29	Manufacturing precision	4, 17, 34, 26	All	1, 32, 35, 23	25, 10	All	26, 2, 18	All	26, 28, 18, 23	10, 18, 32, 39
30	External harm affects the object	All	24, 35, 2	2, 25, 28, 39	35, 10, 2	35, 11, 22, 31	22, 19, 29, 40	22, 19, 29, 40	33, 3, 34	22, 35, 13, 24

(*Continued*)

The Contradiction Matrix (Continued)

		Worsened Feature									
		31	32	33	34	35	36	37	38	39	
	Improved Feature	*Object Generated Harmful Factors*	*Ease of Manufacture*	*Ease of Operation*	*Ease of Repair*	*Adaptability or Versatility*	*Device Complexity*	*Difficulty of Detecting and Measuring*	*Extent of Automation*	*Productivity*	
31	Object-generated harmful factors	All	All	All	All	All	19, 1, 31	2, 21, 27, 1	2	22, 35, 18, 39	
32	Ease of manufacture	All	All	2, 5, 13, 16	35, 1, 11, 9	2, 13, 15	27, 26, 1	6, 28, 11, 1	8, 28, 1	35, 1, 10, 28	
33	Ease of operation	All	2, 5, 12	All	12, 26, 1, 32	15, 34, 1, 16	32, 26, 12, 17	All	1, 34, 12, 3	15, 1, 28	
34	Ease of repair	All	1, 35, 11, 10	1, 12, 26, 15	All	7, 1, 4, 16	35, 1, 13, 11	All	34, 35, 7, 13	1, 32, 10	
35	Adaptability or versatility	All	1, 13, 31	15, 34, 1, 16	1, 16, 7, 4	All	15, 29, 37, 28	All	27, 34, 35	35, 28, 6, 37	
36	Device complexity	19, 1	27, 26, 1, 13	27, 9, 26, 24	1, 13	29, 15, 28, 37	All	15, 10, 37, 28	15, 1, 24	12, 17, 28	
37	Difficulty of detecting and measuring	2, 21	5, 28, 11, 29	2, 5	12, 26	1, 15	15, 10, 37, 28	All	34, 21	35, 18	
38	Extent of automation	2	1, 26, 13	1, 12, 34, 3	1, 35, 13	27, 4, 1, 35	15, 24, 10	34, 27, 25	All	5, 12, 35, 26	
39	Productivity	35, 22, 18, 39	35, 28, 2, 24	1, 28, 7, 10	1, 32, 10, 25	1, 35, 28, 37	12, 17, 28, 24	35, 18, 27, 2	5, 12, 35, 26	All	

Source: Modified from Altshuller, G. S. 1997, *40 Principles*, Worcester, MA: TIC, 135. Used with permission.

Index